T0205539

Studies in Computational Intelligence

Volume 768

Series editor

Janusz Kacprzyk, Polish Academy of Sciences, Warsaw, Poland
e-mail: kacprzyk@ibspan.waw.pl

The series "Studies in Computational Intelligence" (SCI) publishes new developments and advances in the various areas of computational intelligence—quickly and with a high quality. The intent is to cover the theory, applications, and design methods of computational intelligence, as embedded in the fields of engineering, computer science, physics and life sciences, as well as the methodologies behind them. The series contains monographs, lecture notes and edited volumes in computational intelligence spanning the areas of neural networks, connectionist systems, genetic algorithms, evolutionary computation, artificial intelligence, cellular automata, self-organizing systems, soft computing, fuzzy systems, and hybrid intelligent systems. Of particular value to both the contributors and the readership are the short publication timeframe and the world-wide distribution, which enable both wide and rapid dissemination of research output.

More information about this series at http://www.springer.com/series/7092

Saqib Ali · Taiseera Al Balushi
Zia Nadir · Omar Khadeer Hussain

Cyber Security for Cyber Physical Systems

 Springer

Saqib Ali
Department of Information Systems, College
    of Economics and Political Science
Sultan Qaboos University
Al Khoudh, Muscat
Oman

Zia Nadir
Department of Electrical and Computer
    Engineering, College of Engineering
Sultan Qaboos University
Al Khoudh, Muscat
Oman

Taiseera Al Balushi
Department of Information Systems, College
    of Economics and Political Science
Sultan Qaboos University
Al Khoudh, Muscat
Oman

Omar Khadeer Hussain
ADFA
University of New South Wales-UNSW
Canberra, ACT
Australia

ISSN 1860-949X   ISSN 1860-9503 (electronic)
Studies in Computational Intelligence
ISBN 978-3-030-09348-8   ISBN 978-3-319-75880-0 (eBook)
https://doi.org/10.1007/978-3-319-75880-0

Printed on acid-free paper

This Springer imprint is published by Springer Nature
The registered company is Springer International Publishing AG
The registered company address is: Gewerbestrasse 11, 6330 Cham, Switzerland

Preface

This book is designed to be a pioneer yet primary general reference book in cyber-physical systems (CPSs) and their security concerns. It provides foundations which are sound, clear, and comprehensive view of security issues in the domain of CPSs. The book is designed as such it provides the fundamental theoretical background in the area of CPS security and associated domains.

To achieve the aforesaid objectives, the book is divided into eight chapters. Chapter 1 sets the scene by providing a brief overview of CPS, its architecture, components, and the importance of security in it. It also discusses the different types of security attacks and the challenges in addressing them. Chapter 2 discusses the importance of risk in ensuring CPS security. It provides a review of the different risk assessment approaches along with the standards. It then presents a view of how risk management in CPS should be approached for achieving complete resilience in CPS applications. Chapter 3 delves into securing the building blocks of CPS, namely the wireless sensor networks (WSNs). It analyzes the different security issues related to the physical layer, data link layer, network layer, and transport layer. It then highlights the importance of trust and reputation in improving the security in WSN. Chapter 4 details on the mechanisms currently utilized to enhancing WSN security in CPS to deal with both external and internal attacks at different layers. After discussing the different attacks, it provides a comprehensive comparison of the different types of approaches used for the detection and defense of attacks on WSN and CPS. Chapter 5 focusses on industrial control systems (ICSs) and supervisory control and data acquisition (SCADA) systems as they are the critical infrastructures being used in CPS. It discusses the different vulnerabilities of the platform along with the different threats and challenges to provide such CPS communication infrastructure. Chapter 6 studies the interrelationships between embedded systems and CPS. Related to the area of security, it highlights the challenges and gaps with regards to implementing security in CPS. Chapter 7 details on the different aspects of distributed control systems such as its design, architecture, and how they are modeled in CPS. It then discusses the various different security issues that need to be addressed for securing the CPS modeled by using it. Chapter 8 discusses the standards that have been proposed by different organizations to enable seamless operation and inter-compatibility between various different components that are deployed in the CPS environment.

With these topics covered, the book serves as a useful reference book for students in the field of information technology, computer science, or computer engineering who are considering this topic as their further area of study.

Al Khoudh, Muscat, Oman   Saqib Ali
Al Khoudh, Muscat, Oman   Taiseera Al Balushi
Al Khoudh, Muscat, Oman   Zia Nadir
Canberra, Australia   Omar Khadeer Hussain

Acknowledgements

This book would not have been possible without the support of many people. This book is the result of a research project carried out during 2014–2016 at Sultan Qaboos University, Sultanate of Oman. The research leading to these results has received Project Funding from The Research Council of the Sultanate of Oman (TRC) under Research Agreement No [ORG/SQU/ICT/13/011]. The authors would like to thank TRC and Sultan Qaboos University for all the administrative and financial support. We gratefully recognize all the individuals who have supported us throughout this research life and helped us in collecting literature and preparing reports for the project. We would like to thank our research team: Dr. Osama Rehman (Post-Doc), Mohammed Ahsan Kabir Rizvi, Farhan Muhammad, Rana Jacob Jose and Ennan Fatima, and M.Sc. students Ms. Amira Al Zadjali (Risk Management) and Mohammed Al Abri (ICS/SCADA security).

Contents

About the Authors

Dr. Saqib Ali is currently an Associate Professor at the College of Economics and Political Science, Information Systems Department. Dr. Saqib is a self-motivated, capable, and commercially astute IT professional with solid experience across both business and academic settings. He demonstrates proven capabilities in business analysis, software development, information systems, management, and implementation, using this expertise in a commercial environment to introduce innovative strategies and practices that deliver growth and competitive superiority. Dr. Saqib offers an exceptional educational background, maintaining a high profile within the academic arena through continued publishing and actively participating in and leading conferences, workshops, and seminars delivered nationally and internationally. He uses strong communication and interpersonal skills to network successfully, develop strong relationships with clients and colleagues, and deliver engaging and informative course lectures and tutorials that inspire future IT leaders. Corresponding author email: saqib.ali@ieee.org

Dr. Taiseera Al Balushi is currently an Assistant Professor at the College of Economics and Political Science, Information Systems Department. Her main research interests involve the integration of quality concerns into software systems development projects by the application of the widely recognized standards such as ISO, IEEE, etc. Dr. Taiseera is active in the area of teaching and she received Best Teacher Award for the academic year 2012/2013. She is also active in the area of research as she managed to complete successfully two SQU internal grants, two grants from The Research Council of Oman, and one strategic research fund grant from His Majesty Trust Fund of Oman. Dr. Taiseera managed to accomplish a number of industry-related projects with the undergraduate students such as (Royal Oman Police) ROP emergency application and Ministry of Manpower mobile application.

Dr. Zia Nadir is currently an Associate Professor and Assistant Head of ECE department at Sultan Qaboos University. He received B.Sc. (Hons.) in Electrical Engineering with specialization in Communication Systems from University of

Science and Technology, Peshawar Pakistan in 1989. He received D.E.S.S. in microwaves and microelectronics and D.E.A. in Electronics and Ph.D. in Electronics with specialization in Electromagnetic Compatibility from Laboratory of Radio Propagation and Electronics-LRPE (renamed as TELICE), University of Sciences and Technologies, Lille-1 France in 1996, 1997, and 1999, respectively. From 1990 to 1994, he worked in WAPDA, the electricity department of Pakistan. In 1999, he joined the High Voltage and Short Circuit, Laboratory, Islamabad-Pakistan as Assistant Director (Technical) where he was involved mainly in testing procedures, standards, and performing tests on transformers, switchgears, insulators, power transformers, etc. He has won several university and faculty level awards. Since 2001, he is attached with the Electrical and Computer Engineering Department, Sultan Qaboos University Muscat, Oman.

Dr. Omar Khadeer Hussain is currently a Senior Lecturer at the School of Business, UNSW Canberra. Prior to joining the School in February 2014 as a Lecturer, he worked as a Senior Research Fellow at Curtin University. Omar's research areas of interests are Logistics and Supply Chain Management, Distributed and Grid Systems, Decision Support, Group Support Systems and their applications to Logistics areas. Omar's research has been published in international journals such as *The Computer Journal, Journal of Intelligent Manufacturing*, etc. He has won university and faculty level awards for his research and as the main supervisor has supervised seven Ph.D. students to completion. In 2011, he was awarded with an APDI Fellowship from the ARC on a Linkage project with Prof. Elizabeth Chang as the lead CI.

Abbreviations

ACK	Acknowledgement
ACL	Access control lists
AES	Advanced encryption standard
ANSI	American National Standard Institute
ARM	Advanced RISC machines
ARP	Address resolution protocol
B-MAC	Berkeley media access control
BS	British Standards
CCTV	Closed circuit television
CERT	Computer Emergency Readiness Team
CIA	Confidentiality Integrity Availability
CIP	Critical infrastructure protection
COBIT	Control objective for information and related technologies
COSO	Committee of Sponsoring Organization of the Tread-away Commission
CPSs	Cyber-physical systems
CPU	Central processing unit
CRC	Cyclic redundancy check
CSA	Canadian Standards Association
CTS	Clear to send
DCSs	Distributed control systems
DEFP	Different forking probabilities
DES	Data encryption standard
DNP	Distributed network protocol
DoS	Denial of service
DRBTS	Distributed reputation based trust system
DS	Dansk Standards
DSDVV	Destination sequenced distance vector protocol
DSSS	Direct sequence spread spectrum
EDP	Economic dispatch problem
EFP	Enforced fractal propagation

ETSI	European Telecommunication Standards Institute
EU	European Union
EVM	Electronic voltmeter
FEAL	Fast encryption algorithm
FHSS	Frequency hopping spread spectrum
GAO	Government Accountability Office
GCC	Gulf Cooperation Council
GETAR	Geographic, energy and trust aware routing
GSM	Global system for mobile communications
GSO	GCC Standardization Organization
GTMS	Ground truth measurement system
HMI	Human–machine interface
HVAC	Heating ventilation and air conditioning
IBC	ID-based cryptography
ICSs	Industrial control systems
ICT	Information communication technology
IDS	Intrusion detection system
IEC	International Electrotechnical Commission
IED	Intelligent electronic devices
IEEE	Institute of Electrical and Electronic Engineers
IoT	Internet of Things
IP	Internet protocol
IPC	Association Connecting Electronics Industries
ISACA	Information Systems Audit and Control Association
ISMS	Information security management system
ISO	International Organization for Standardization
IT	Information technology
ITIL	IT infrastructure library
ITU	International Telecommunication Union
JAID	Jamming avoidance itinerary design
LDTS	Lightweight and dependable trust system
L-MAC	Lightweight medium access control
LTM	Lightweight trust model
M2M	Machine to machine
MAC	Medium access control
MANET	Mobile ad hoc network
MPR	Multi parent routing
NEN	Netherlands Standardization Institute
NERC	North American Electric Reliability Corporation
NIST	National Institute of Standards and Technology
NISTIR	National Institute of Standards and Technology Internal Report
OLE	Object linking and embedding
OPC	OLE for process control
OS	Operating system
OT	Operational technology

PALS	Physically asynchronous but logically synchronous
PLC	Programmable logic controllers
PTZ	Pan tilt zoom
QoS	Quality of service
RFID	Radio frequency identification
RSSI	Received signal strength indicator
RTS	Request to send
RTU	Remote terminal unit
SAE	International Society of Automotive Engineers
SANS	SysAdmin, Audit, Networking and Security Institute Standards
SC	Smart grid
SCADA	Supervisory control and data acquisition
S-MAC	Sensor medium access control
SNARE	Sensor node attached reputation evaluator
SQL	Structured query language
TDM	Time division multiplexing
TRM	Trust and reputation management
UK	United Kingdom
UML	Unified Modeling Language
UPS	Uninterruptable power supply
USA	United States of America
VLA	Virtual force localization algorithm
VLAN	Virtual local area network
WSAN	Wireless sensor and actuator network
WSN	Wireless sensor network

List of Figures

List of Tables

Chapter 1
Cyber-Physical Systems Security

In recent times, Cyber-Physical Systems (CPS) have become an emerging paradigm for controlling and managing the ever-growing number of cyber-connected devices and has seen an increased use in different applications. Among the various benefits that CPS provides leading to its wide adaptation, one of them is its ability to seamlessly integrate systems from cyber and physical domain to create value. However, to fully realize such benefits various complementary aspects needs to be considered. One such aspect is security, which secures the communication between different devices and ensures the prevention of unauthorized access. This chapter sets the base for the rest of the book by discussing the structure of the CPS and its different associated components. It then discusses some of the various different attacks from which CPS need to be secured and the challenges involved in addressing them.

1.1 Introduction

A CPS is known as the new generation of mix systems consisting of computational and physical capabilities. These systems interact with humans through different models and subsystems. A CPS aims to monitor the physical processes behaviors and then take actions to make the physical environment work properly and better (Wang et al. 2010). CPSs are also known as the systems that have tight integrations among the computation, networking, and physical objects. The embedded devices in CPS systems are also networked to sense, manage, and monitor all the physical components (Xia et al. 2011). The importance of CPS is growing rapidly because of its design of smooth integration of computational and physical components and also because it uses clear interface standards for a smooth communication among its channels (Xia et al. 2011). It is generally used in critical national infrastructure, for example, electric energy power, water distribution systems, petroleum systems, and so on.

© Springer International Publishing AG 2018
S. Ali et al., *Cyber Security for Cyber Physical Systems*, Studies in Computational Intelligence 768, https://doi.org/10.1007/978-3-319-75880-0_1

The technologies of the CPS are becoming increasingly important as they guarantee to provide a comfortable, healthy, secure, and safe environment. To improve their functionalities and operations, most critical infrastructures are now being prepared with the modern CPS components and they have started to use cyber-based services (O'Reilly 2013).

Unlike the traditional computer and network systems, CPSs are complex "systems of systems", where vulnerability and risks can have new and distributed impacts that can result in a catastrophic failure of critical infrastructure and services. When combining the physical systems with cyber systems in CPS, different impacts of threats will be carried out. Subsequently, new risk assessment methodologies are needed to be applied in the CPS to safeguard them against such threats (O'Reilly 2013).

This chapter will provide readers with an appropriate and a critical theoretical background that would help to understand the CPSs' concepts, theories, and applications in order to comprehend its cybersecurity issues involved in them.

1.2   Cyber-Physical Systems

Cyber-physical systems have become important enablers in improving the living conditions across the world and has been applied and implemented in various critical areas which are related to the national economy (Rajkumar et al. 2010). For example such systems have been used extensively in improving the health and welfare of individuals in remote and rural places. Some main and critical examples include the power grid, next generation systems of automobiles, intelligent highways, next-generation systems of air vehicles, and airspace management (Dong et al. 2015). CPS is an engineered system that can change the way to interact with physical systems or devices, similar to Internet revolution. Moreover, CPS can be defined as the integration of computation and physical processes and it is therefore not the merging of both physical and the cyber parts.

CPS is a tight integration among the computation, networking, and physical objects, where the embedded devices are networked in a way to sense, manage, and monitor the physical world (Xia et al. 2011). Although there is no unified definition for CPS; generally it is defined as the new generation of systems that mixes computational and physical capabilities to be able to interact with humans through different models. It also aims to monitor the physical processes behavior and take proper actions to make the physical environment work properly (Wang et al. 2010).

Although the integration of both the physical processes and computing devices is considered to be new, the term "embedded systems" was used previously in explaining and describing the engineered systems. However, embedded systems are considered to be as closed "boxes" that do not use the computing capabilities outside. For example, some of these popular applications are communication systems; games, aircraft control systems, and toys (Lee 2008). Researchers envision that these CPSs will exceed today's systems in terms of flexibility, efficiency, functionality, dependability, security, and usability. The great advantages of the CPS in future are their quick response, precision, and adaptability to work in risky conditions, distributed coordination, high efficiency, and enhancement for the societal well-being (Axelrod 2013).

Universal applications and services can be enabled by the CPS to expressively improve the quality. Moreover, it has been used to make the applications, operations, and services more effective and efficient. Examples of CPS may include unified medical systems, transportation, automated traffic control systems, energy conservation, critical infrastructure, environmental control, etc. (Xia et al. 2011). Moreover, capacities of systems such as information processing, real-time communication, component independency, and physical objects interaction with the networked environment can be improved by using the CPS (Zhang et al. 2013).

1.3 CPS Architecture and Components

CPS has been presented essentially as a two-tier structure, the physical and computing part. Main activities, which come under the physical parts, are to sense the physical environment, data collection, and then executing the computing part decisions. The computing parts mainly analyze and process the physical part data and finally make the decision. Such feedback will control the relation of the two parts. Although these two parts (cyber and physical) are essentially different, however, they are integrated and combined in a way that affects each other by information (Hu et al. 2012).

Usually the physical process in CPS is monitored and controlled by the other cyber system. Cyber system is a networked system of a number of small devices with some other capabilities such as sensing, computing, and communication (Wang et al. 2010). In its later descriptions, CPS was presented as a three-tier architecture (La and Kim 2010). Those tiers consisted of an environmental, service, and control trier. **An Environmental Tier**: It consists of physical devices along with the end users, who use the devices and their associated physical environment. **Service Tier**: It consists of a typical computing environment with some services like Service-Oriented Architecture (SOA). **Control Tier**: This tier receives a monitored data gathered from the sensors, which helps to make controlling decisions. It also helps to find the right services with the help of service framework (Hu et al. 2012). Figure 1.1 depicts a general overview of CPS conceptual layout with associated system components.

According to the systems theory, a CPS merges the physical space with the cyberspace by integrating the computing and communication capabilities with monitoring and controlling information between physical world entities. This will build a bridge between the cyberspace and the physical space as illustrated by

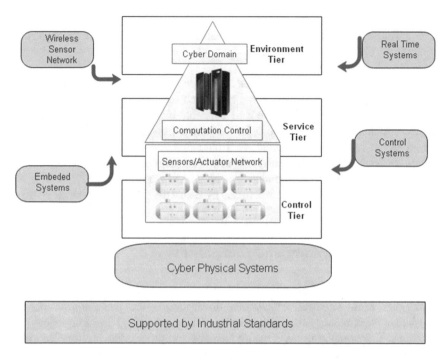

Fig. 1.1 Cyber-physical systems conceptual layout

(Dong et al. 2015). Generally, each physical component has a cyber capability in the CPS. A CPS behavior is a full integration of the computational and physical action, where computing is deeply embedded in all physical components (Lee 2006).

1.4 CPS Technology and Security Importance

Many benefits are exposed by merging of computing and communication process with the physical process in a CPS. These benefits include a more safe and efficient physical systems; minimizing the building and operating costs of these physical systems; and providing new capabilities from the individual machines by working together (Kaiyu et al. 2010).

In general, CPS systems are mixed and more heterogeneous consisting of multiple types of physical systems and communication and computation models. A CPS design is much more than the combination and union of both computational and physical systems and many effective methods are available for designing such systems (Kaiyu et al. 2010). Some typical CPS applications are health and traffic monitoring system, process control system, energy, and environmental monitoring systems (Wan et al. 2010b). Most of these applications require a high level of assurance because of the safety requirements (Lee 2008).

The range of devices in a CPS environment may include anything from the most simple hardware, which are sensors to all high-level and complex computers, and cloud that manages all systems data and controls them. A wide range of reliable and compromised networks moves the information and commands from one place to another in such system. Therefore, a CPS must be reliable and secure, as it is one of its main requirements (Madden 2012). The system is complex in a CPS environment since both strict and relaxed operating rules are available in its hardware and software. For interacting with the physical space, a way to translate the information to a computational value is required. In general there are four distinct categories of systems that create a CPS (Madden 2012): They are: **Sensors**—a device that translates the physical world to the digital world, **Embedded System**—a standalone device with hardware, software, and mechanical parts which do not require a special thing to operate, however may not be completely useful, **High-Level System**—can be a desktop computer or a high-end super computer, which can perform mass quantities of computations, advanced logic, and vast multitude of functions at the same time and **Physical World**—where the system exists and includes physics and real-time constraints. To understand the physical world more, a sensor ID is added to an embedded system or to a high-level system. Sensors are simple devices that require something to read in order to become a useful device. In order for the high-level system to perform, it constantly interacts with the physical world and works on the input provided by the sensors. With this, the CPS gets another critical problem to address which is of the information passing through a communication channel. Since CPS requires many networks protocols to operate, communication is not restricted to specific types of networks. Therefore, evaluating and securing the communication channels in the CPS operations is critical (Wan et al. 2010a). Security of a CPS includes all the areas of analyzing the computer, physical part, and all the interactions between them. However, there is a lack of security analysis that covers both aspects. Both physical and computer components need to be considered for the security analysis (Akella et al. 2010). Due to the CPS's complexity, many properties are required to take a considerable amount of designing and planning to ensure that the CPS meets the required capabilities to operate (Madden 2012).

1.5 CPS Security and Cyberattacks

The most critical challenge in CPS is to keep using the new emerging technologies such as cloud and social computing in connecting the different devices, while at the same time manage the different types of risks for such technology. Unfortunately, such risks are also one of the least understood challenges in CPS (Sha et al. 2009). Security in the CPS is relatively a new area and can be defined as the ability to secure the data and the operational capabilities of the systems from unauthorized access (Banerjee et al. 2012). In a CPS, protection of the information is difficult and therefore security is a critical and complex task. This is due to the nature of the architecture design where data, communication, processing, and communication channels are combined (Madden 2012).

Confidentiality, integrity, availability, and authenticity are the basic properties of a computer and CPSs' security. Protection of the information refers to the confidentiality and data validity. The correctness is the integrity, and the capability of using any resource is the availability (Habash et al. 2013), where authenticity ensures the communication, data exchange, and parties involved are genuine and unaffected. CPS security objectives are as follows.

Confidentiality This refers to the system's capability to prevent and avoid disclosure of information to the outside unauthorized users or system (Han et al. 2007). In a CPS, confidentiality is required and necessary but it alone is not sufficient to maintain the privacy (Pham et al. 2010).

Integrity Integrity refers to the insurance of any data or resources from any modifications without proper authorization. In a CPS domain, achieving integrity is the key goal through its preventing, detecting, and deterring deception attacks on the communication between sensors and the actuators or controller (Madden et al. 2010).

Availability This refers that the system should be available to serve its purpose. This relates to all of the CPSs' aspects such as processes that are used to store and send information etc. The physical parts are used to perform physical process, and communication channels are used to access these physical parts (Wang et al. 2010). CPS, through high availability usually aims to provide service and at the same time prevent computing, control, communication from corruptions due to different reasons like hardware failures, system upgrades, power outages, or denial-of-service attacks (Work et al. 2008).

Authenticity This ensures that the data, transaction, and communications are genuine. Availability is also important and required for the validation of every party involved (Stallings 2006). In a CPS, realizing the authenticity in all the processes such as sensing, communications, and actuations is one of the main tasks for the processing of the system (Wang et al. 2010).

CPSs are exposed to different types of attacks and the security priorities have to be defined. Some of the common attacks that impact on the above-mentioned priorities of the CPS are as follows (Wang et al. 2010).

Eavesdropping In this attack, an attacker can interrupt information communicated by the systems without interfering in the working of the system. It is also called as a passive attack in which the attacker will only observe the system's operation. User's privacy will also be violated in this type of attack.

Compromised Key attack In this attack, the compromised key is held by an attacker, which would help in gaining access to the secured communication. Both the sender and receiver will not be aware of this activity by the attacker. Then any modification can be done using the compromised key on the captured data.

Man-in-the-Middle attack In this attack, false messages are sent to the recipient and they may result in false negative or false positive actions. The recipient can take action when it is not required or the recipient may think everything is fine and does not take an action when it is required.

Denial-of-Service attack This attack is a network attack that prevents legitimate requests from the network resources. In this type of attack normal processes are disrupted and the attacker can gain access over the system from flood controller to shut down. The denial-of-service attack prevents the normal work or use of the system. In this type of attack, cybercriminals can attack any computer from anywhere. They may not target to harm the systems; however, negative side effects may be caused like infecting the systems with malware. Moreover, the cyberattacks are developed by the physical attacks. They are cheaper, less risky, and are easier to replicate by the attacker (Cardenas et al. 2009). In the recent years, CPS' vulnerabilities analysis of the external attacks got more attention (Pasqualetti 2012).

Cybersecurity is a very critical concern that is predicted to be among the fastest growing segments of the Information Technology (IT) sector. There is a huge growth to be prevented from such attacks. This is evidenced by the huge investments from the companies, who are trying to secure their computing environments (Watts 2003). Till now, CPS' security issues related to power networks, linear networks, and water networks have received a considerable attention in the literature (Pasqualetti 2012).

1.6 Challenges in CPS Security

CPS security is quite a new area and generally can be defined as the ability of the systems to allow only authorized access to its data and operational capabilities. Because of the newness in this area, current efforts in CPS security are focused on mapping the solutions from existing fields (Venkatasubramanian et al. 2009). However, these are not beneficial vulnerabilities of CPS, since these issues are different and securing such systems is a hard and challenging task. However, the criticality to address the security issues in CPS is huge due to the mission-critical nature of the CPS applications that will have profound consequences, if they are not addressed (Halperin et al. 2008). Another factor to be considered for CPS security is its information sensitivity and detail about the physical process (Banerjee et al. 2012).

Since most CPS systems are not designed for security but for the core functionality, therefore a cybersecurity requirement is not one of the concerns that the CPS manufacturers include in their original product development. However, as explained earlier CPSs are exposed to many cybersecurity breaches. One of the main reasons is the long lifetime of the CPS systems and the rapid changes in security mechanisms during their lifetime. Another factor is connecting to the Internet exposes the CPS to different types of exploits and exposes the private communications happening between different devices. Attackers take advantage of the open remote connection for the support and operations purposes. In addition to this, another factor is the continuous lack of personnel training in security area. Some CPSs' suppliers may no longer exist and that would be another source of vulnerabilities (O'Reilly 2013).

Cardenas et al. (2009), discusses the three key challenges in securing CPSs. In general, having an understanding of the threats in CPS is a challenge to be addressed and to know the possible consequences of attacks. Identifying the unique properties of the CPSs and their differences from traditional IT security is a critical challenge in this area. Finally, discussing and defining the current security mechanisms, which are applicable to cyber-physical systems, are limited.

1.7 Conclusion

Operational Technology (OT) uses the traditional IT technology over the CPSs and embedded advanced technologies to control, manage, and monitor various critical operational and processing tasks, such as the function of controlling the voltage line sensors and voltage circuit breakers in the smart grid environment. Moreover, the OT depends mainly on the sensors, actuators, PLC, RTU, ICS SCADA systems, industrial servers, and ICT infrastructure to fasten the whole OT's operations collectively. Although many organizations have isolated the OT from IT for security purposes, the new trend is still growing to converge both networks, which, as a result, raise an enormous of technical and security challenges (Proctor 2016). For secured integration, the various security precautions should be taken to protect corporate networks, such as building continuously systematic updated information about the status of the IT network and the security gap which should be filled. Also, a great care should be taken to monitor and control the whole IT network's infrastructure, devices vendors, contractors and models, location, the basic configuration of firewalls, routers, switches, servers, computers, printers, Ethernet cabling and ports, and wireless access points (Ogbu and Oksiuk 2016).

The composition of this book is to discuss the status of cybersecurity of each system components of CPSs. The book is structured as follows: Chapter 2 discusses the importance of risk management for CPSs' security. Chapters 3 and 4 provide extensive literature review and security mechanisms for Wireless Sensor Networks used in CPS. Chapter 5 presents Industrial Control Systems SCADA systems' security for CPS. Chapters 6 and 7 present Embedded and Distributed control systems security for CPS. At the end, Chap. 8 presents the analysis for standards being employed for CPS.

References

Akella, R., Tang, H., & McMillin, B. M. (2010). Analysis of information flow security in cyber-physical systems. *International Journal of Critical Infrastructure Protection, 3,* 157–173.

Axelrod, C. W. (2013). Managing the risks of cyber-physical systems. In *Systems, applications and technology conference (LISAT), 2013 IEEE Long Island* (pp. 1–6). IEEE.

Banerjee, A., Venkatasubramanian, K. K., Mukherjee, T., & Gupta, S. K. S. (2012). Ensuring safety, security, and sustainability of mission-critical cyber-physical systems. *Proceedings of the IEEE, 100,* 283–299.

Cardenas, A., Amin, S., Sinopoli, B., Giani, A., Perrig, A. & Sastry, S. (2009). Challenges for securing cyber physical systems. In *Workshop on Future Directions in Cyber-Physical Systems Security.*

Dong, P., Han, Y., Guo, X., & Xie, F. (2015). A systematic review of studies on cyber physical system security. *International Journal of Security and Its Applications, 9,* 155–164.

Habash, R. W., Groza, V., & Burr, K. (2013). Risk management framework for the power grid cyber-physical security. *British Journal of Applied Science & Technology, 3,* 1070.

Halperin, D., Heydt-Benjamin, T. S., Fu, K., Kohno, T., & Maisel, W. H. (2008). Security and privacy for implantable medical devices. In *IEEE Pervasive Computing,* Vol. 7.

Han, J., Shah, A., Luk, M., & Perrig, A. (2007). Don't sweat your privacy. In *Proceedings of 5th International Workshop on Privacy in UbiComp (UbiPriv'07).*

Hu, L., Xie, N., Kuang, Z., & Zhao, K. (2012). Review of cyber-physical system architecture. In *Object/Component/Service-Oriented Real-Time Distributed Computing Workshops (ISORCW), 2012 15th IEEE International Symposium on, 2012.* IEEE, pp. 25–30.

Kaiyu, W., Man, K., & Hughes, J. (2010). Towards a unified framework for cyber-physical systems. In *Proceedings of the 1st International Symposium on Cryptography,* pp. 292–295.

La, H. J., & Kim, S. D. (2010). A service-based approach to designing cyber physical systems. In *Computer and Information Science (ICIS), 2010 IEEE/ACIS 9th International Conference on.* IEEE, pp. 895–900.

Lee, E. A. (2006). Cyber-physical systems-are computing foundations adequate. In *Position Paper for NSF Workshop on Cyber-Physical Systems: Research Motivation, Techniques and Roadmap.*

Lee, E. A. (2008). Cyber physical systems: Design challenges. In *Object Oriented Real-Time Distributed Computing (ISORC), 2008 11th IEEE International Symposium on.* IEEE, pp. 363–369.

Madden, J. (2012). *Security analysis of a cyber physical system: A car example.* Missouri University of Science and Technology.

Madden, J., Mcmillin, B., & Sinha, A. (2010). Environmental obfuscation of a cyber physical system-vehicle example. In *Computer Software and Applications Conference Workshops (COMPSACW), 2010 IEEE 34th Annual.* IEEE, pp. 176–181.

O'Reilly, P. (2013). Designed-in cybersecurity for cyber-physical systems.

Ogbu, J. O., & Oksiuk, A. (2016). Information protection of data processing center against cyber attacks. IEEE, pp. 132–134, 1509029788.

Pasqualetti, F. (2012). *Secure control systems: A control-theoretic approach to cyber-physical security.* Santa Barbara: University of California.

Pham, N., Abdelzaher, T., & Nath, S. (2010). On bounding data stream privacy in distributed cyber-physical systems. In *Sensor Networks, Ubiquitous, and Trustworthy Computing (SUTC), 2010 IEEE International Conference on.* IEEE, pp. 221–228.

Proctor, B. (2016). *Operational technology: OT is the blood brother of IT* [Online]. Missouri: Malisko Engineering, Inc. Available: http://www.malisko.com/operational-technology/. Accessed 2016.

Rajkumar, R. R., Lee, I., Sha, L., & Stankovic, J. (2010). Cyber-physical systems: The next computing revolution. In *Proceedings of the 47th Design Automation Conference*. ACM, pp. 731–736.

Sha, L., Gopalakrishnan, S., Liu, X., Wang, Q., Yu, P. S., & Tsai, J. J. (2009). Machine learning in cyber trust: Security, privacy, and reliability.

Stallings, W. (2006). *Cryptography and network security: Principles and practices*. Pearson Education India.

Venkatasubramanian, K. K., Banerjee, A., & Gupta, S. K. (2009). Green and sustainable cyber-physical security solutions for body area networks. In *Wearable and Implantable Body Sensor Networks, 2009. BSN 2009. Sixth International Workshop on*. IEEE, pp. 240–245.

Wan, K., Hughes, D., Man, K. L., & Krilavičius, T. (2010a). Composition challenges and approaches for cyber physical systems. In *Networked Embedded Systems for Enterprise Applications (NESEA), 2010 IEEE International Conference on*. IEEE, pp. 1–7.

Wan, K., Man, K., & Hughes, D. (2010b). Specification, analyzing challenges and approaches for cyber-physical systems (CPS). *Engineering Letters, 18*.

Wang, E. K., Ye, Y., Xu, X., Yiu, S.-M., Hui, L. C. K., & Chow, K.-P. (2010). Security issues and challenges for cyber physical system. In *Proceedings of the 2010 IEEE/ACM Int'l Conference on Green Computing and Communications & Int'l Conference on Cyber, Physical and Social Computing*. IEEE Computer Society, pp. 733–738.

Watts, D. (2003). Security and vulnerability in electric power systems. In *35th North American Power Symposium,* pp. 559–566.

Work, D., Bayen, A., & Jacobson, Q. (2008). Automotive cyber physical systems in the context of human mobility. In *National Workshop on High-Confidence Automotive Cyber-Physical Systems*, pp. 3–4.

Xia, F., Vinel, A., Gao, R., Wang, L., & Qiu, T. (2011). Evaluating IEEE 802.15. 4 for cyber-physical systems. *EURASIP Journal on Wireless Communications and Networking, 2011,* 596397.

Zhang, L., Qing, W., & Bin, T. (2013). Security threats and measures for the cyber-physical systems. *The Journal of China Universities of Posts and Telecommunications, 20,* 25–29.

Chapter 2
Risk Management for CPS Security

As cyber-physical systems combine physical systems with the cyber domain, to safeguard the communication medium and address the growing security issues, a well-designed risk management is required. The available risk assessment approaches in the area of cybersecurity may not be applied directly to CPS since they are different in many aspects. This chapter explores, reviews, and analyzes risk assessment approaches and frameworks recommended for CPS risk management is presented. It then proposes a reference style framework for enhancing cybersecurity to ensure having complete resilience of the CPS architecture.

2.1 Introduction

In the previous chapter, security as one of the main factors that need to be managed to achieve the expected benefits in its different CPS areas of applications is high-lighted. Due to the nature of CPS, security needs to be implemented at two different levels, namely, the physical and cyber level. Physical level security aims to secure the components of CPS at the environmental tier. In other words, such level of security aims to secure the physical devices and the environment in which they are in from unauthorized access and/or control. Solutions to achieve such type of security are broadly related to physically securing the required area. On the other hand, cyber level security too aims to secure and prevent the unauthorized use of physical devices. But the channel from where such threats arise is not only from the physical medium but also from the cyber one. With the proliferation of digital communities and communications, the security threats from the cyber domain, also known as cybersecurity has become one of the most important aspects to be addressed. This is because the threats from such level affect and impact the service and control tiers of the CPS architecture, thus making them vulnerable and impacting some of their basic properties in communication such as confidentiality, integrity, availability, authenticity of communication, etc.

© Springer International Publishing AG 2018
S. Ali et al., *Cyber Security for Cyber Physical Systems*, Studies in Computational Intelligence 768, https://doi.org/10.1007/978-3-319-75880-0_2

In the recent past, there have been a number of cyberattacks which have impacted the affected parties both financially (Charleston 2017) and reputationally (Manuel 2015). Researchers in the literature have attempted to address such security issues by identifying the different vulnerabilities (Humayed et al. 2017), the different types of attacks that are possible, both in the cyber and physical sphere (Humayed et al. 2017; Giraldo et al. 2017; Shafi 2012; Wu et al. 2016) and proposing approaches to mitigate them (Wardell et al. 2016) through different methods. While having such knowledge, methods and techniques are important and beneficial; from a cybersecurity perspective, it is also important to understand that having a one size fits all approach is not possible and workable (Madhyastha 2017). Furthermore, it is also possible that with a constant increase in cyberattacks, no two cyberattacks may be the same. They may have different traits which are purposely developed to create maximum disruptions in CPS operations. Hence, in such cases, the previously developed and captured knowledge of cyberattacks may not be the only information and knowledge required to prevent attacks and make the CPS secure.

2.2 Risk—Its Definition and the Two Different Ways by Which It Is Managed

The term risk broadly signifies the occurrence of events that will lead the activity toward an unwanted outcome. Even though various researchers, for example, Waters (Waters 2011) argue that risk should not be always be associated with a negative outcome, but because of its history of being used in the field of insurance, risk is most of the times seen as a factor to minimize and manage. Taking the broad definition of risk proposed in Hussain et al. (2013), in the context of this book, risk broadly is defined as a possibility of the intended outcomes of the CPS applications not being achieved due to possible breaches in the applied security measures. Hence, it is important for such events that lead to risk to be identified and managed.

There are two broad approaches in the literature that have been proposed to identify and manage risk (Hussain et al., 2013). They are either adopting a reactive or proactive approach. Consistent with its terminology, a reactive approach reacts to address and manage situations of risks after they occur. On the contrary, a proactive approach, rather than waiting for risk events to occur first before managing them, predetermines what events of risk may have an impact on CPS security and accordingly develops plans to address them. The difference between these two types of approaches is as follows:

(a) The reactive approach does not stop the security of the CPS application being impacted by risk events but the proactive approach does. This implies that the reactive approach does not stop the CPS application from experiencing disruptions, loss of revenue, loss of reputation, etc., but the proactive approach, if done successfully has the ability to do so.

(b) As the reactive approach attempts to address CPS security threats after they occur, it requires more of problem-solving techniques. On the other hand, as the proactive approach attempts to address CPS security threats before they occur, it needs the ability to predetermine possible problems and techniques to avoid them.

2.3 How Does the Above Discussion Relate to How CPS Security Should Be Managed?

As mentioned earlier, the area of security in CPS has been extremely well researched in the literature. Ashibani and Mahmoud (2017) summarize different CPS security issues in its three layers of perception, transmission, and application and propose possible solutions that can be applied at single- and multilayers. Maple (2017) summarizes the different challenges related to security and privacy in IoT in different applications such as autonomous vehicles, health, well-being, etc., and discusses on who should be responsible and accountable for this. Wang et al. (2010) discuss the major types of attacks to CPS and propose a context-aware security framework for improving the overall security. The proposed approach utilizes context as the basis on which the decision whether to allow a user access to the CPS application is made. Seifert and Reza (2016) in the context of a healthcare application, tests the applicability of two architectures, namely, publish and subscribe and blackboard to ascertain which one performs better against some of the commonly known security threats. Machado et al. (2016) propose an approach of preserving the intellectual property of CPS system by controlling the software embedded in the CPS devises that manage them.

While the above mentioned works along with others in the literature are important, it can be argued that the security threats against which they aim to secure the CPS platform and architecture are known. In other words, the proposed approaches aim to develop resilience of CPS applications against threats that are already well known and identified. While this improves the capability of the CPS architecture to improve its resilience against known threats, it does not provide any assistance in improving the CPS architecture security against threats that are not currently known. Referring to the discussion from the previous section, such an approach of managing CPS security has similarities with the reactive process of risk management as it can only react to new events of security threats after they occur. However, they may not be useful if the resilience of the CPS architecture needs to be improved from security threats that may not be that well known. In the next section, we discuss some of the existing risk management approaches utilized in CPS applications.

2.4   Concept and Importance of Risk Management

Risk management is important and plays a critical role in many fields, for example, economics, biology and operations (Djemame et al. 2014). It is a process of balancing the operational and economic costs and allows the IT managers to protect the IT systems and data of their organization. Risk management process is not for IT operations only, it also helps making the decision in all areas of the organization's functions (Stoneburner et al. 2002). To simplify, a risk management is a process of identifying risk, assessing risk, and taking decision on risk-based issues to reduce risks to an acceptable level. Its objective is to support the overall vision and mission of the organization and to enable the organization in accomplishing its mission(s). In general, the risk management covers three main processes: risk assessment, risk mitigation, and evaluation and assessment (Stoneburner et al. 2002).

As discussed earlier, while defining risk, most definitions focus on the uncertainty of outcomes and differ in characterizing these outcomes. Mainly a risk is described as having adverse consequences whereas others seem to be natural. In general, a risk refers to the uncertainty of future events and is defined as the expression of the likelihood of occurrence and the impact even may lead to achieve the organization's objective. In IT, a risk is defined as an appearance of the asset value, the vulnerability to that risk and the threat exists in the organization. In other words, a risk is the possibility of a risky or dangerous event to occur which will affect or impact achieving objectives. Risk analysis is required to help in making decisions about organization's objectives. Two calculations are necessary for each risk: risk likelihood or probability and the impact or consequences (Berg 2010; Djemame et al. 2014).

For any risk management, the significant concepts to be addressed are as follows:

- an asset which is worth value and requires protection,
- an unwanted incident that affects or reduces the value of an asset,
- a threat which is the possible cause of an unwanted incident,
- a vulnerability which is a weakness, error, or deficiency that opens for, or may be exploited by, a threat to cause harm to or reduce the value of an asset, and
- Finally, the risk is defined as the likelihood of an unwanted incident and incident consequence on a particular asset.

To explain the above risk concepts, for example, a server is an asset; computer will be affected by virus which is considered to be a threat for that computer and virus protection software which is not up to date is considered to be vulnerable. This will lead to an unwanted incident, which is an access for a hacker on this server. The likelihood of a backdoor on the server to be created by the virus may be medium, but the consequence to harm the server would be high (Djemame et al. 2014). In general, there is an essential issue in characterizing any risk and that is because of appropriately carrying out the following steps in the risk management process (Djemame et al. 2014):

- analyzing the triggering event and formulating an accurate structure of the risk,
- estimating the loss related to every event, once a risk is realized, and
- estimating the possibilities of an event by using statistical methods or subjective judgments.

After identifying the possible risks, which is a first step in the risk management process, the risks are then assessed by their potential severity of loss and possibility of occurrence. The process of assessing is called as risk assessment. In risk management process, the risks are measured or assessed and then managed by developing strategies and controlling their consequences. While managing risks, the set of actions are defined which are required to manage a specific risk and the actions are applied as a strategy, which consists of the following measures (Hillson 2002):

- transferring the risk to another party to manage the risk better,
- avoiding the risk and eliminating the uncertainty to occur,
- reducing the effects of the risk, and
- accepting the risk and limiting the consequences of a specific risk.

Many risks have occurred in the IT systems as well; therefore, IT managers must ensure that the organization has the required capabilities to achieve its missions. They should identify the security capabilities that will enable their IT systems to provide the desired level of mission support in order to face the threats and risks. Defining a well-structured risk management methodology and to use it effectively can help the organization management to find suitable controls for specific threats and risks (Stoneburner et al. 2002).

Nowadays, cyber-physical systems are being increasingly seen in most critical areas; with smart grid is one of the key examples. This development leads to many risk issues to appear because of the cyber-physical system's design. When systems are developed from a Cyber-Physical System's perspective, the risks of the whole system will be greater than the sum of individual component's risks. The assessment and mitigation of these overall risks require the understanding of threats to the data-processing systems and the damage that results from it (Axelrod 2013). Managing the risk of Internet and information-rich system is difficult because of the cybersecurity issues that result from the interaction of cyber and physical process in the CPS. In addition, the mission-critical nature of cyber-physical systems makes them more vulnerable to different attacks (Fletcher and Liu 2011).

The process of risk assessment helps in evaluating what needs to be protected. It is an ongoing process of evaluating the risks and then to establish an appropriate risk management program for the organization. In information security, for example, the risk management program should be appropriate for the degree of risk related to the organization's systems, networks, and information assets. Basically, the process of identifying threats to the information or information systems, determining the likelihood of occurrence of the threat, and identifying system vulnerabilities that could be exploited by the threat is defined as the risk assessment (Stoneburner et al. 2002).

2.4.1   Risk Assessment

A risk is the degree of possible wrong that may occur in an established process, and risk assessment is one of the key activities in risk management process. The risk assessment process is the first phase of risk management. Assessing risk means identifying the threats, and then determining the likelihood and impact (Tiwana and Keil 2004). There are five phases involved in the risk assessment process: identification, analysis, evaluation, control and mitigation, and documentation (Kumsuprom et al. 2008). In general, an IT risk assessment helps the organizations in implementing new business changes and use the information systems to get the appropriate changes. At the same time, these implementations of IT systems have also exposed to IT-related risks such as strategic risk, financial risk, operational risk and technological risk (Kumsuprom et al. 2008). Therefore, IT risk assessment polices and strategies have been developed to minimize such risks.

Much of the early analysis of cyber-threats and cybersecurity appears to have "The Sky is Falling" theme (Lewis 2002). "Information security—the safeguarding of computer systems and the integrity, confidentiality, and availability of the data they contain—has long been recognized as a critical national policy issue" (Cashell et al. 2004). At the same time, its importance is growing more in different aspects of the most modern countries.

The risk assessment's concept has been presented and discussed as a general methodology and or a specific type of risk in many areas such as grids and clouds (Djemame et al. 2014). However, the chances of the severity of future risks to appear could be more than what has been seen today of the cyberattacks and information breaches. This is because of the dramatically increase in the number of such cyberattacks in the world (Cashell et al. 2004). These cyberattacks, network breaches, and information breaches can create complex risks which could affect the new areas for national security and public policy (Lewis 2002).

2.4.2   Risk Assessment Approaches for CPS

A number of standards, guidelines, and recommendations address the cybersecurity area and risk assessment in particular. At the same time, some existing risk assessment approaches are available and have been developed to mitigate the cybersecurity risks. Basically, some approaches were introduced and some were researched by the scholars and they came up with some guidelines. In Table 2.1, some of these approaches are listed; however, it should be noted that the table is not comprehensive.

Different risk assessment approaches are available in the area of CPS, and they may obtain different results for the similar risks. In the assessment process, the CPS characteristics must be taken into considerations to find a suitable and appropriate assessment method. For example, an attack tree risk assessment method includes good features that make it suitable to develop the CPS risk assessment (Yong et al. 2013).

Table 2.1 Risk assessment approaches for CPS

Approach/Article	Description
Physical and cyber risk analysis tool (PACRAT) 2013 (Macdonald et al. 2013)	It includes cyber-physical information about facility layout, network topology, and emplaced safeguards to detect, delay, and respond to attacks, to identify the pathways most vulnerable to attack, and to evaluate how often the safeguards are compromised for a given threat or an adversary type
Collaborative security management system (CYSM system), 2014 (Karantjias et al. 2014)	This approach introduced a collaborative security management system named as (CYSM system), which enables ports' operators to analyse and manage: (a) model physical and cyber assets and interdependencies; (b) internal/external/ interdependent physical and cyber threats/ vulnerabilities; and (c) physical and cyber risks against the requirements specified in the ISPS Code and ISO27001.
Cyber-physical system risk assessment 2013 (Yong et al. 2013)	It is three-level CPS architecture. It sums up the traditional risk assessment methods, analyzes the differences between cyber-physical system security and traditional IT system security, and proposes a risk assessment idea for the CPS
Risk management 2013 (Axelrod 2013)	The suggested mitigation strategies include substantial improvements in designing, developing, and testing the current systems, and a process for ensuring that the combined systems meet equally strict security and safety requirements
Goal-based assessment (Merrell et al. 2010)	The assurance cases help in addressing these problems. Viewing an assessment approach in terms of an assurance case clarifies the underlying motivation for the assessment and supports more rigorous analysis. The paper also shows how the assurance case method has been used to guide the development of an assessment approach called the cyber resilience review (CRR)
A method for modeling and evaluation of the security of cyber-physical systems (Orojloo and Azgomi 2014)	It considers those cyberattacks that can lead to physical damages. The factors impacting attacker's decision-making in the process of cyberattack to cyber-physical system are also taken into account. Furthermore, for describing the attacker and the system's behaviors over time, the uniform probability distributions are used in a state-based semi-Markov chain (SMC) model. The security analysis is carried out from mean time to security failure (MTTSF), steady-state security, and steady-state physical availability

(continued)

Table 2.1 (continued)

Approach/Article	Description
Risk-based approach (Habash et al. 2013b)	The paper evaluates the security threats to the smart grid as well as the health risks of smart meters currently under implementation in the context of radio frequency (RF) radiation. Finally, a combined framework for risk management in major technological and health domains has been proposed
Implementing an integrated security management framework to ensure a secure smart grid (Enose 2014)	This paper, therefore, introduces an integrated security management framework that offers a critical infrastructure-grade security, to multiple utility technologies in establishing an enterprise-wide integrated security management system. This comprehensive security architecture offers an improved interconnection of the diverse systems and establishes both physical security and cybersecurity, integrated to all functional aspects of the grid
The behavioral approach (Enose 2014)	The behavior-based anomaly/misuse detection procedures, described herein, have resulted in a system diagnostic technology capable of timely detection, identification, and prediction of hardware malfunction caused by cyber attacks on its computer control systems. The approach is implemented with a numerically efficient system call processing algorithms for an off-line functionality extraction and an online functionality matching
Cyber-physical vulnerability assessment for power-grid infrastructures (Vellaithurai et al. 2015)	Cyber-physical security indices to measure the security level of the underlying cyber-physical setting. CPINDEX installs appropriate cyber-side instrumentation and probes on the individual host systems to dynamically capture and profile low-level system activities such as inter-process communications among operating system assets. CPINDEX uses the generated logs along with the topological information about the power network configuration to build stochastic Bayesian network models of the whole cyber-physical infrastructure and update them dynamically based on the current state of the underlying power system. Finally, CPINDEX implements belief propagation algorithms on the created stochastic models combined with a novel graph-theoretic power system indexing algorithm to calculate the cyber-physical index, i.e., to measure the security level of the system's current cyber-physical state

(continued)

Table 2.1 (continued)

Approach/Article	Description
Impact assessment approach 2014 (Charitoudi and Blyth 2014)	It focuses on the responsibilities and the dependencies that flow through the supply chain, mapping them down into an agent-based socio-technical model. This method should be especially useful for modeling consequences across all levels of complex organizations' networks business processes, business roles, and systems

2.4.3 Smart Grid Risk Assessment

Smart grid risk can be defined as the probability of threats that may use vulnerability to cause harm to a computer, network, and system. This will then result in operational and business consequences for the smart grid. In a smart grid, the complex functionality of threats and their consequences will make the cybersecurity risk difficult to measure (Clements et al. 2011).

On the other hand, the smart grid cybersecurity risk assessment's main objective is to identify different vulnerabilities and threat risks and then determine their impact. Ultimately, the resulted outcome of the risk assessment should be used to define the security requirements and controls (Hecht et al. 2014).

While defining the cyber risk management process for grid systems, it is important to decide the amount to be spent on information security and how the budget is allocated for it. Although there is a clear gap in the data and the measurement of capabilities, businesses and government, where strong risk management programs can be developed by organizations (Cashell et al. 2004). Many methodologies are in use nowadays to qualitatively measure the risk instead of measuring it quantitatively. A common practice is to rank information assets according to (a) how valuable they are and (b) how vulnerable they are to attack (Cashell et al. 2004). It is important to focus on the combination of both power application security and supporting infrastructure security when forming or developing the risk assessment process for power-grid systems (Sridhar et al. 2012).

It is important to highlight and present the best practice checklist for the smart grid risk management process. The checklist below summarizes security's best practices and controls which are presented by the National Rural Electric Cooperative Association (NRECA) in implementing risk management (Lebanidze 2011):

- provide active executive sponsorship,
- assign responsibility for security risk management to a senior manager,
- define the system,
- identify and classify critical cyber assets,
- identify and analyze the electronic security perimeter(s) (ESPs),

- perform a vulnerability assessment,
- assess risks to system information and assets,
- select security controls, and
- monitor and assess the effectiveness of controls.

Generally, as CPS is facing many different kinds of risks, a well-designed risk assessment for the CPS will help in providing general view of its security status and support allocation of such resources; therefore, it is necessary to take in consideration four elements which are as follows: asset, threat, vulnerability, and damage (Lu et al. 2013). Moreover, many comprehensive surveys and studies of risk assessment methods are done in the area of cybersecurity, and a little number of frameworks has been identified in the area of risk assessment for critical (energy) infrastructures.

The guidelines for smart grid cybersecurity, which are set from high-level recommendations and standards applicable to smart grid architecture for the US, have been developed by NIST (NIST-IR 7628) (Group 2010). However, they do not provide a general approach for assessing cybersecurity risks. At the same time, NIST-IR 7628 and ISO 27002 were used as basis for a report on smart grid security by ENISA providing a set of specific security measures and meant to establish a minimum level of cybersecurity. In the smart grid environment, the risk assessment should be performed during the system life cycle. Several factors must be taken account when defining the risk assessment process such as the size of the organization, the implementation cost of the measures, etc. The risk assessment allows defining a threshold for the minimum acceptance level of any risk (ENISA 2012).

Some risk assessment programs and guidelines have been published or are currently under development to provide the smart grid organizations with a good reference to implement such programs in their organization. A number of smart grid risk assessment methods and efforts are presented in Table 2.2, which can be used to mitigate cybersecurity risks. However, almost all of them focus on addressing threats and evaluating risks against specific technical requirements. Some available risk assessment results are used for suggesting the system design and others are used in very specific types of failures in the smart grid component.

The CPS cyber risk management is defined as the program that is used to manage cybersecurity risks of organization's operations. Effective risk management and assessment programs must enable the organization management to achieve the objectives defined in their strategy by proper decision-making. Organizations need to identify its critical assets at the first stage and then assess the related risks to these assets. In addition, the risk management process is a continuous process which includes the assessment process to manage and evaluate the risks. Key decision about the risks of such environment must be sent to the management level for proper actions. A good presentation of these risks is necessary for managers to take decision.

The process of the risk management mainly includes assigning priorities to risks and proper budget to be established to measure and implement such process which

Table 2.2 Smart grid assessment

Approach	Description
Unified risk management 2010 (Ray et al. 2010)	It helps evaluate risks to adopting protection measures enhances security and reliability in smart grid information exchange system
Risk-based approach 2013 (Habash et al. 2013a)	It is a combined risk management, which coordinates assessment of cyber and power-grid risks
BBN-based approach 2013 (Brezhnev and Kharchenko 2013)	The Bayesian belief network (BBN) where nodes represent different CSGS and NPP; different types of influences cause links and consider critical substation with a critical load
Transient stability assessment (TSA) (Cepeda et al. 2011)	Multichannel singular spectrum analysis (MSSA), principal component analysis (PCA), and support vector machine classifier (SVM-C) tools are included in it. It finds hidden patterns in electric signals, and SVM-C, and can use those patterns for efficiently classifying the system vulnerability status
Game-theoretic approach 2015 (Law et al. 2015)	It provides an estimate of the defender's loss due to load shedding in simulated scenarios, calculated risks incorporated into a stochastic security game model as input parameters. The decisions on defensive measures are obtained by solving the set using dynamic programming techniques that take into account resource constraints. Thus, the formulated security game provides an analytical framework for choosing the best response strategies against attackers and minimizing the potential risks
Probabilistic approaches 2014 (Ciapessoni et al. 2014)	The probabilistic power flow is combined with copulas to represent an effective way the correlation information among variables. It is effective in determining uncertainties in operating condition
Physical and cyber risk analysis tool (PACRAT) 2013 (Macdonald et al. 2013)	It gives information about the facility layout, network topology, and emplaced safeguards to evaluate the facilities ability to detect, delay, and respond to attacks, to identify those pathways which are most vulnerable to attack. The vulnerability assessment is about cyber-physical interdependencies
Stochastic programming 2012 (Noyan 2012)	It is the impact of load curtailments within a required short response time. This approach can build agility into utilizing customers' energy decisions, thereby helping exploit the energy demand response capability and achieve a strategic advantage over competitors

(continued)

Table 2.2 (continued)

Approach	Description
Value at risk (VaR) 2012 (Ahmadi-Javid 2012)	Value at risk (VaR) is proposed as a measure of system security and is applied to an illustrative six-bus test system. It is the security assessment in operation planning horizon
EMS and SCADA 2012 (Liu et al. 2012)	It detects, locates, and determines compromises in a SCADA system, and autonomously remediates the affected components, within an appropriate time frame to minimize the negative impact on the function and power. It provides guidelines and examples by which power utility companies and control centers can provide their own self-assessments
MILP formulation 2015 (Ceseña et al. 2015)	It is an original generalized flow diagram to map the energy flow from a wide range of different DMG systems. Moreover, it allows synthetic writing of relevant equations. A new generalized MILP formulation for the optimization of DMG systems captures operational flexibility to respond to energy price signals by exploiting multifactor arbitrage opportunities in the presence of different DMG technologies and time arbitrage opportunities in the presence of thermal storage (in case), and a stochastic programming model, again, is formulated as MILP, which allows type, size, and investment time optimization of melt energy multicomponent flexible planning schemes considering long-term uncertainty and intrinsic risk hedging
Reliable implementation of robust adaptive topology control (Bopp et al. 2014)	It indicates how relay settings may be changed due to switching actions; it also provides an online algorithm for detection of relay misoperations, which identifies the lines that may be switched back to service after being tripped erroneously by a relay
Protection performance assessment (Bopp et al. 2014)	The in-depth evaluation allows the pinpointing of false settings in individual relays and for various operating and fault conditions. If false or improvable settings are detected, then the solution supports the calculation of new improved protection settings To enhance the quality control, a new set of settings can be validated and verified through simulation before application in order to maximize network utilization and grid reliability by reducing the risk of unwanted protection and action. It demonstrates how the applied solution can help to assure and improve the quality of protection settings through event-triggered or regular application

(continued)

Table 2.2 (continued)

Approach	Description
Multi-elements and multi-dimensions risk evaluation 2012 (Qiang et al. 2012)	In this paper, first, different risks of smart grid can be identified from five aspects, including financial risk, security risk, technical risk, management risk, and policy risk. These basic risk factors almost can cover any fields of smart grid. Second, based on risk elements identified, a multidimensions risk assessment matrix method is developed to evaluate the risk of smart grid considering the time and the regional development nature of local smart grid
Risk analysis and probabilistic survivability assessment (RAPSA) (Taylor et al. 2002)	It is an assessment approach for power substation hardening. It is a new cybersecurity assessment approach, which merges survivability system analysis (SSA) with probability risk assessment (PRA). The method adds quantitative information to the process oriented SSA method, which assists in decision-making among security options

is one of the most important high management responsibilities. In this respect, it is critical to follow a risk assessment approach that will provide a conceder to the management at first stage.

2.5 Adopting Security Standards

To this effect, organizations are spending considerable resources in building proper information security risk management programs that would eventually address the risks they are exposed to. These programs need to be established on solid foundations which are the reason why enterprises look for standards and frameworks that are widely accepted and common across organizations.

The NIST's cybersecurity framework for the cybersecurity may not constitute a foolproof formula that includes leading practices framework because it relinquishes or chooses to defer the implementation of the "Voluntary Guidelines" which can be remembered by those implemented and proved to be successful, and also quickly adopt the cybersecurity for organizations that extend well beyond regulatory and legal advantages and those different parameters can be saved from the body. The framework owns or operates in the organizations' critically targeted infrastructure, while in fact, the adoption proves beneficial for businesses in virtually all industries (Al-Ahmad and Mohammad 2012).

The standards, in general, are meant to provide uniformity that would ease the understanding and management of the concerned areas. The businesses find themselves in need to adopt standards for various reasons which vary from business requirements to regulators and compliance mandates. The establishments of proper corporate governance, increasing risk awareness and competing with other enterprises are some business drivers mentioned. Some firms pursue certifications to meet the market expectations and improve their marketing image. A major business driver for standards' adoption is filling in the gaps and lack of experience in certain areas where firms are not able to build or establish proprietary standards based on their staff competencies (Al-Ahmad and Mohammad 2012).

2.6 Security Frameworks

2.6.1 ISO 27000 Set

ISO 27001 (formally ISO/IEC 27001), an information security management system (ISMS) is a specification. An ISMS is an organization's information risk management process which includes all the legal, physical, and technical controls that include a framework of policies and procedures (Al-Ahmad and Mohammad 2012).

According to the documentation, ISO 27001 is about "installation for implementing, operating, monitoring, reviewing, maintaining and improving an Information Security Management System to provide a model" (Almorsy et al. 2011). ISO 27001 is a top-down, risk-based approach and the technology is neutral.

27000 series standards and the individual will be populated with a series of documents. A number of these documents are already well known, and in fact, have been published. The final number and the other publishing details are yet to be

Table 2.3 ISO standard series

ISO27001 This is the specification for an information security management system (an ISMS) which has replaced the old BS7799-2 standard	ISO27002 This is the 27000 series standard number of what was originally the ISO 17799 standard (which itself was formerly known as BS7799-1)
ISO27003 This will be the official number of a new standard intended to offer guidance for the implementation of an ISMS (IS management system)	ISO27004 This standard covers information security system management measurement and metrics, including suggested ISO27002 aligned controls
ISO27005 This is the methodology independent ISO standard for information security risk management	ISO27006 This standard provides guidelines for the accreditation of organizations offering ISMS certification

determined and scheduled for publication. The following matrix shown in Table 2.3 reflects the currently known position for the major operational standards in the series (Humphreys 2006):

2.6.2 IT Infrastructure Library (ITIL 3.0)

Another view of ITIL V3, service strategy, and service design volume of this service is based and focused on the definition. Their goal is a new service, and the improvement of an existing one is based on the design of the matter, not the design and development of IT services.

A service design is purely based on the technology, but the business and technical environment addresses the masses in order to negotiate a solution, so the planned service support is required to take into consideration, the entire supply chain (Marrone and Kolbe 2011).

2.6.3 NIST Risk Management Framework (RMF)

It is an information system for the selection and specification of security controls— that is, the organizational risk management that covers an organization. A wide information security program has been completed as part of the organizations or individuals associated with the operation of information system risk offers.

The selection and specification of the risk-based approach to security control effectiveness, efficiency, and applicable laws, directives, executive orders, regulations, standards, or regulations because of the constraints understandable (also known as risk management framework) for the following activities related to organizational risk management are paramount to an effective information security program and the system development life cycle and the federal enterprise architecture within the context of both new and legacy information systems that can be applied.

2.6.4 OCTAVE Set

Operationally, critical threat, asset, and vulnerability assessment (OCTAVE) level of risk and planning defenses against cyberattacks is determined by a security framework. Framework organizations are likely to help reduce the risk of threats to determine the likely outcome of an attack is successful and defines a method for dealing with attacks. In OCTAVE, people within the organization get leverage, and so, the experience and expertise is built. The first step that they pose is to build threats based on risk profiles. The process to conduct a risk assessment, relevant to the organization, goes on (Alberts et al. 1999).

OCTAVE defines three phases:

Phase 1: Asset-based risk profiles.
Phase 2: Identify infrastructure vulnerabilities.
Phase 3: Develop security strategy and plans.

OCTAVE for the United States Department of Defense, Carnegie Mellon University (CMU) was developed in 2001. Since then, the structure has gone through several evolutionary stages, but the basic principles and goals have remained the same. There are two versions: OCTAVE-S, flat hierarchical structures, a simple method for small organizations, and OCTAVE Allegro, larger organizations or those with multilayered structure which is a more comprehensive version (Alberts et al. 1999).

2.6.5 COSO Framework

The Committee of Sponsoring Organizations of the Tread way Commission (COSO) has published an Internal Control-Integrated framework in 1992. The joint initiative aims to provide guidance on enterprise risk management, internal control, and fraud deterrence, and this is through the development of frameworks. The goal of COSO is to provide leadership by the development of comprehensive frameworks and guidelines to improve the overall performance of the organization (Spira and Page 2003).

In 2013, COSO has issued an update of the internal control framework in order to reflect the changes in the business world. The new 2013 framework retains the five components of internal control which are control environment, risk assessment, information and communication, control activities, and monitoring activities.

Four principles have been introduced in particular, which are related to risk assessment. These principles are principle 6, 7, 8, and 9:

- 6: The organization specifies objectives with sufficient clarity in order to enable the identification and assessment of the risks relating to objectives.
- 7: The organization identifies risks for the achievement of its objectives.
- 8: The organization considers the potential for fraud in assessing risks for the achievement of objectives.
- 9: The organization identifies and assesses changes that could significantly impact the system of internal control.

In year 2017, COSO update its framework known as enterprise risk management framework or integrated framework. They updated 2004 framework by integrating and aligning their components with strategy and performance, emphasis on the inclusion of risk in both the setting up their strategies and in driving performances.

This update addresses the evolution of enterprise risk management and need for organizations to improve their approach toward risk management in order to meet the dynamic or evolving business environment (Accountants 2004).

2.7 Analyzing Risk Management Frameworks

Over the years, several risk frameworks have been developed and each has its advantages and disadvantages. Basically, they all call for a multidisciplinary team. The discipline requires the organizational assets, inventory risks, define, control, evaluation, and risk, ending up with an estimate of a magnitude.

Perhaps the most well-known risk framework OCTAVE comes in three sizes. The original is a full-featured version for large organizations with sufficient documentation. OCTAVE-S multidisciplinary group can be represented by fewer people, where small groups, sometimes with business knowledge in particular, are designed for technical ones. Documentation burden is low and lighter weight in the process. OCTAVE's latest product in the series is more of a trivial texture and has a more focused approach than its predecessors, which leads to the Allegro. The Allegro at the start of the process requires extra discipline, assets for the awareness, and visual systems, applications, and environments as containers. The abstract assessment of the scope of information (such as protected health information), and the identity and information is stored, processed, or transmitted, in which the container is needed to assess the risk across. One of the benefits of OCTAVE series worksheets is to document every step in the process providing templates for each of the arrangements. These are either directly or intended to be used for a particular organization.

As described in the NIST Special Publication 800-30 NIST framework, it can be applied to any property that is a normal one. It uses a slightly different terminology than the OCTAVE, but follows a similar structure. Its brevity and more solid components (e.g., system) focus on risk assessment making it a good candidate for the new organizations. In addition, it is defined by NIST that the government agencies and organizations that work with them are approved for use.

ISACA's COBIT, ISO 27001 and 27002 are risk management programs for organizations that require management and security arrangements. Both offer, but do not need their own version of the risk management frameworks: COBIT and ISO 27005. They recommend repeatable methodologies of risk assessment that should take place when specified. COBIT included in all of the ISO 27000 series is designed to deal with security; consequently, risk assessments required by each of those consistent areas. In other words, COBIT risk assessment as described in the risk goes beyond safety, and ISO 27005, in particular, is focused on safety, development, business continuity, and IT operations to include other types of risks.

ISO 27005 NIST follows a similar structure, but defines the conditions differently. For instance, infrastructure threats, vulnerabilities, and controls are considered in the context of the establishment called the move, including risk identification and assessment, and a risk analysis that discusses the steps and documents potential risk and business impact. ISO 27005 includes annexes with other risk frameworks, assessment which are relevant to a particular business, and apply to quantify risk in ways that depend on the organization.

Different types of organizations are continuously exposed to many kinds of risks. For example, information technology risks, people risks and process risks. Frameworks and standards are available in order to help the organizations manage such risks. However, they need to select the most adequate ones for their organization and business and use them to address these security risks. The Committee of Sponsoring Organizations of the Tread way Commission (COSO) framework is one of the available frameworks that provide risk assessment process within its risk management processes.

The COSO emphasizes upon the development of a framework that will fully integrate the management of risk into the organization. The framework assures that the corporate-wide process is supportive, iterative and effective. It means that the risk management will be an active component in governance, strategy and planning, management, reporting processes, policies, values, and culture. The framework provides for the integration of risk management, reporting and accountability. It intends to adapt to the particular needs and structure of each organization.

2.8 How Should Complete Resilience of CPS Architecture Should Look like in Terms of its Security?

Existing approaches adopt a silo-based view of the threats which impact CPS security while aiming to increase the resilience of the CPS architecture. In other words, they consider an inside-out approach while maintaining the security of the CPS architecture. However, apart from that, there may be other threats (outside-in) which are not currently known, but when they occur will have the ability to disrupt the flow of information in the cyber domain. This is logical as new security advisories constantly come out and if complete resilience against security attacks in CPS architecture needs to be developed, appropriate methods and techniques need to be developed that can identify such threats and take mitigation actions accordingly. In other words, there is a need to be proactive against in the management of CPS security to improve the resilience of the CPS architecture.

The importance of being proactive in areas such as risk management has been stressed in the literature in many different domains. However, existing methods in CPS security aim to secure the architecture based on the current known security threats. Apart from doing that, it is recommended that there needs to a constant analysis of new threats that could possibly arise and develop actions to mitigate them. With the huge amount of information being generated these days, one way by which this can be done is to combine existing CPS security approaches with data science techniques that will enable CPS security managers in the better management of the CPS goals. Existing research in the literature (Wang et al. 2016) emphasizes on the need to use big data analytics in applications such as logistics and supply chain. But such approaches focus on utilizing predictive, prescriptive, and descriptive processes. While these approaches will provide insights for the better management of operations, they will not be helpful in the domain of security management unless the analysis is conducted on a real-time basis. This is because one of the main objectives of security management is to take a proactive approach that analyzes and manages events that may have a detrimental impact on the ultimate goal. To achieve this, real-time stream processing of information is required by security managers which can then be combined with the traditional existing approaches for security management of CPS systems that are currently in use.

2.9 Reference Style for Security Risk Management in Cyber-Physical Systems

The modern Web with its fascinating range of tools has become the "laboratory" for understanding the real-time processing of events. Further to that, a range of techniques has been proposed to mine this information and make sense of *large social and information networks*. These techniques fall within the research area of social and information network analysis (SINA), in which link analysis, network community detection, diffusion, and information propagation on the Web are some of the main themes. Using such models to analyze the data allows CPS security managers to understand behaviors and patterns, and to distinguish information of interest from expected phenomena from which insights can be generated. Numerous data science techniques are used widely in the literature, such as network graph analysis, information capture and analysis on a real-time basis, enterprise knowledge graph, sentiment analysis, and social media analysis. However, not all techniques will be beneficial for every scenario. The challenge here is to identify which techniques combined with which analysis are the most beneficial according to the required objective of the analysis, and to use it accordingly.

To achieve the above aim, it is not only important to have the required data but also to acquire the data in the required format and at the required time. This falls into the category of data representation, data management, data access, and the timely processing of information. Appropriate techniques need to be developed that will enable CPS security managers to utilize the data in the short time frame in which it is beneficial for managing supply chain risk. Not all data on the Web is reliable and trustworthy. A key research issue, therefore, is to study the provenance of data to ascertain its reliability and trustworthiness before utilizing it in the analysis.

2.10 Conclusion

With the increased need of the proper security of the CPS infrastructure from cyberattacks, a number and diversity of risk assessment methodologies have risen to help achieving this goal. Many international standards and guidelines from different parties on risk assessment and risk management frameworks can confuse a company that is seeking security. However, proper guidelines and standards can be used to develop an appropriate risk assessment for the CPS and fulfill the security requirements for such critical systems.

In this chapter, a literature review was done in the area of CPS risk assessment in order to provide the basic information regarding the available studies and effort in this area. Furthermore, a list of the most relevant risk assessment approaches for CPS and mainly for smart grid has been presented. The risk assessment process helps the organizations in determining the assets that are at risk and to help in defining the controls to mitigate those risks. The risk assessment process is mainly based on the threats and vulnerabilities that could cause damage to any critical system.

An overall risk assessment and risk management framework helps to properly evaluate the importance of security issues within an organization. Although a quite good number of frameworks that can be used to perform risk assessments are available, however, a number of proper and suitable assessment approaches that can be used in conjunction with these frameworks are very low. However, at the same time, it is important to understand that for complete resilience against CPS security risks we need to be proactive in identifying the different adversaries that may occur and manage them accordingly.

References

Accountants, A. I. O. C. P. (2004). COSO enterprise risk management—Integrated framework. Available online: https://www.cpa2biz.com/AST/Main/CPA2BIZ_Primary/InternalControls/COSO/PRDOVR ~ PC-990015/PC-990015.jsp.

Ahmadi-Javid, A. (2012). Entropic value-at-risk: A new coherent risk measure. *Journal of Optimization Theory and Applications,* pp. 1–19.

Al-Ahmad, W., & Mohammad, B. (2012). Can a single security framework address information security risks adequately? *International Journal of Digital Information and Wireless Communications, 2,* 222–230.

Alberts, C. J., Behrens, S. G., Pethia, R. D., & Wilson, W. R. (1999). *Operationally critical threat, asset, and vulnerability evaluation (OCTAVE) framework* (Vol. 1).

Almorsy, M., Grundy, J., & Ibrahim, A. S. (2011). Collaboration-based cloud computing security management framework. In *International Conference on Cloud Computing.* IEEE.

Ashibani, Y., & Mahmoud, Q. H. (2017). Cyber physical systems security: Analysis, challenges and solutions. *Computers & Security, 68,* 81–97.

Axelrod, C. W. (2013). Managing the risks of cyber-physical systems. In *Long Island Conference on Systems, Applications and Technology.* IEEE.

Berg, H.-P. (2010). Risk management: Procedures, methods and experiences. *Risk Management, 1,* 79–95.

Bopp, T., Ganjavi, R., Krebs, R., Ntsin, B., Dauer, M., & Jaeger, J. (2014). Improving grid reliability through application of protection security assessment. In *IET international conference on developments in power system protection.* Copenhagen: IET.

Brezhnev, E., & Kharchenko, V. (2013). BBN-based approach for assessment of smart grid and nuclear power plant interaction. In *East-west design & test symposium.* Rostov-on-Don: IEEE.

Cashell, B., Jackson, W. D., Jickling, M., & Webel, B. (2004). The economic impact of cyber-attacks. Available online: http://www.au.af.mil/au/awc/awcgate/crs/rl32331.pdf.

Cepeda, J., Colomé, D., & Castrillón, N. (2011). Dynamic vulnerability assessment due to transient instability based on data mining analysis for smart grid applications. In *2011 IEEE PES conference on innovative smart grid technologies (ISGT Latin America).* IEEE.

Ceseña, E. A. M., Capuder, T., & Mancarella, P. (2015). Flexible distributed multienergy generation system expansion planning under uncertainty. *IEEE Transactions on Smart Grid,* p. 1.

Charitoudi, K., & Blyth, A. J. (2014). An agent-based socio-technical approach to impact assessment for cyber defense. *Information Security Journal: A Global Perspective, 23,* 125–136.

Charleston, L. J. (2017). Three of the biggest cyber security threats to Australian business. Available: http://www.huffingtonpost.com.au/2017/04/05/three-of-the-biggest-cyber-security-threats-to-australian-busine_a_22027681/. Accessed April 6, 2017.

Ciapessoni, E., Cirio, D., Pitto, A., Massucco, S., & Silvestro, F. (2014). A novel approach to account for uncertainty and correlations in probabilistic power flow. In *Innovative smart grid technologies conference Europe (ISGT-Europe), 2014 IEEE PES.* IEEE.

Clements, S. L., Kirkham, H., Elizondo, M., & Lu, S. (2011). Protecting the smart grid: A risk based approach. In *Power and Energy Society general meeting, 2011 IEEE.* IEEE.

Djemame, K., Armstrong, D., Guitart, J., & Macias, M. (2014). A risk assessment framework for cloud computing. *IEEE Transactions on Cloud Computing, 1.*

ENISA. (2012). *Annex II. Security aspects of the smart grid.* Heraklion: European Network and Information Security Agency.

Enose, N. (2014). Implementing an integrated security management framework to ensure a secure smart grid. In *2014 International conference on advances in computing, communications and informatics (ICACCI).* IEEE.

Fletcher, K. K., & Liu, X. F. (2011). Security requirements analysis, specification, prioritization and policy development in cyber-physical systems. In *2011 5th International conference on secure software integration and reliability improvement companion (SSIRI-C).* IEEE.

Giraldo, J., Sarkar, E., Cardenas, A. A., Maniatakos, M., & Kantarcioglu, M. (2017). Security and privacy in cyber-physical systems: A survey of surveys. *IEEE Design & Test, 34,* 7–17.

Group, S. G. I. P. C. S. W. (2010). *NISTIR 7628-guidelines for smart grid cyber security.*

Habash, R. W., Groza, V., & Burr, K. (2013a). Risk management framework for the power grid cyber-physical security. *British Journal of Applied Science & Technology, 3,* 1070–1085.

Habash, R. W., Groza, V., Krewski, D., & Paoli, G. (2013b). A risk assessment framework for the smart grid. In *2013 IEEE conference on electrical power & energy conference (EPEC).* IEEE.

Hecht, T., Langer, L., & Smith, P. (2014). Cybersecurity risk assessment in smart grids. *Tagungsband ComForEn 2014,* 39.

Hillson, D. (2002). Extending the risk process to manage opportunities. *International Journal of Project Management, 20,* 235–240.

Humayed, A., Lin, J., Li F., & Luo, B. (2017). Cyber-physical systems security—a survey. https://ARXIV.ORG/ABS/1701.04525.

Humphreys, T. (2006). State-of-the-art information security management systems with ISO/IEC 27001: 2005. *ISO Management Systems, 6,* 1.

Hussain, O. K., Dillon, T. S., Hussain, F. K., & Chang, E. J. (2013). *Risk Assessment and management in the networked economy.* Berlin Heidelberg: Springer.

Karantjias, A., Polemi, N., & Papastergiou, S. (2014). Advanced security management system for critical infrastructures. In *IISA 2014, The 5th international conference on information, intelligence, systems and applications*. IEEE.

Kumsuprom, S., Corbitt, B., & Pittayachawan, S. (2008). ICT risk management in organizations: Case studies in Thai business. In *19th Australasian conference on information system*. Christchurch: ACIS.

Law, Y. W., Alpcan, T., & Palaniswami, M. (2015). Security games for risk minimization in automatic generation control. *IEEE Transactions on Power Systems, 30,* 223–232.

Lebanidze, E. (2011). *Guide to developing a cyber security and risk mitigation plan*. Arlington, VA: National Rural Electric Cooperative Association.

Lewis, J. A. (2002). Assessing the risks of cyber terrorism, cyber war and other cyber threats. *Center for Strategic and International Studies(CSIS)*, 1–12.

Liu, C.-C., Stefanov, A., Hong, J., & Panciatici, P. (2012). Intruders in the grid. *IEEE Power and Energy magazine, 10,* 58–66, 1540–7977.

Lu, T., Xu, B., Guo, X., Zhao, L., & Xie, F. (2013). A new multilevel framework for cyber-physical system security. In *First international workshop on the swarm at the edge of the cloud*. Montreal: TerraSwarm.

Macdonald, D., Clements, S. L., Patrick, S. W., Perkins, C., Muller, G., Lancaster, M. J., & Hutton, W. (2013). Cyber/physical security vulnerability assessment integration. In *2013 IEEE PES Innovative Smart Grid Technologies (ISGT)* (pp. 1–6). Washington, D.C.: IEEE.

Machado, R. C., Boccardo, D. R., De Sá, V. G. P.D., & Szwarcfiter, J. L. (2016). Software control and intellectual property protection in cyber-physical systems. In *EURASIP Journal on Information Security, 2016,* 8.

Madhyastha, S. (2017). Cyber security—One size does not fil all. *Cyber security by design*. Available online: https://www.stickman.com.au/cyber-security-one-size-not-fit-all/. Accessed April 3, 2017.

Manuel, D. (2015, October 29). The reputational damage of data breaches: don't hope for customer apathy. *CSO Bloogers*. Available online from: https://www.cso.com.au/blog/cso-bloggers/2015/10/29/the-reputational-damage-of-data-breaches-dont-hope-for-customer-apathy/.

Maple, C. (2017). Security and privacy in the internet of things. *Journal of Cyber Policy, 2,* 155–184.

Marrone, M., & Kolbe, L. M. (2011). Impact of IT service management frameworks on the IT organization. *Business & Information Systems Engineering, 3,* 5–18.

Merrell, S., Moore, A. P., & Stevens, J. F. (2010). Goal-based assessment for the cybersecurity of critical infrastructure. In *IEEE International Conference on Technologies for Homeland Security (HST)* (pp. 84–88). Waltham, MA. IEEE.

Noyan, N. (2012). Risk-averse two-stage stochastic programming with an application to disaster management. *Computers & Operations Research, 39,* 541–559, 0305–0548.

Orojloo, H., & Azgomi, M. A. (2014). A method for modeling and evaluation of the security of cyber-physical systems. In *2014 11th International ISC conference on Information Security and Cryptology (ISCISC)* (pp 131–136). Tehran: IEEE.

Qiang, S., Yibin, Z., Dong, H., Zheng, Y., & Jianwei, Z. (2012). Multi-elements and multi-dimensions risk evaluation of smart grid. In *2012 IEEE conference on innovative smart grid technologies-Asia (ISGT Asia)* (pp. 1–6). IEEE: Tianjin.

Ray, P. D., Harnoor, R., & Hentea, M. (2010). Smart power grid security: A unified risk management approach. In *2010 IEEE International Carnahan conference on security technology (ICCST)* (pp. 276–285). San Jose, CA: IEEE.

Seifert, D., & Reza, H. (2016). A security analysis of cyber-physical systems architecture for healthcare. *Computers, 27,* 1–24.

Shafi, Q. (2012). Cyber physical systems security: A brief survey. In *12th International Conference on Computational Science and Its Applications* (pp. 146–150), June 18–21, 2012.

Spira, L. F., & Page, M. (2003). Risk management: The reinvention of internal control and the changing role of internal audit. *Accounting, Auditing & Accountability Journal, 16,* 640–661.

Sridhar, S., Hahn, A., & Govindarasu, M. (2012). Cyber–physical system security for the electric power grid. *Proceedings of the IEEE, 100,* 210–224.

Stoneburner, G., Goguen, A., & Feringa, A. (2002). Risk management guide for information technology systems. In *Nist special publication 800-30* (pp. 2–56).

Taylor, C., Krings, A., & Alves-Foss, J. (2002). Risk analysis and probabilistic survivability assessment (RAPSA): An assessment approach for power substation hardening. In *ACM Workshop on scientific aspects of cyber terrorism,* Washington D.C.

Tiwana, A., & Keil, M. (2004). The one-minute risk assessment tool. *Communications of the ACM, 47,* 73–77.

Vellaithurai, C., Srivastava, A., Zonouz, S., & Berthier, R. (2015). CPINDEX: Cyber-physical vulnerability assessment for power-grid infrastructures. *IEEE Transactions on Smart Grid, 6,* 566–575.

Waters, D. (2011). *Supply chain risk management—Vulnerability and resilience in logistics.* Great Britain: Kogan Page.

Wang, G., Gunasekaran, A., Ngai, E. W. T., & Papadopoulos, T. (2016). Big data analytics in logistics and supply chain management: Certain investigations for research and applications. *International Journal of Production Economics, 176,* 98–110.

Wang, E. K., Ye, Y., Xu, X., Yiu, S. M., Hui, L. C. K., & Chow, K. P. (2010). Security issues and challenges for cyber physical system. green computing and communications (GreenCom). In *2010 IEEE/ACM International conference on & international conference on cyber, physical and social computing (CPSCom)* (pp. 733–738), December 18–20, 2010.

Wardell, D. C., Mills, R. F., Peterson, G. L., & Oxley, M. E. (2016). A method for revealing and addressing security vulnerabilities in cyber-physical systems by modeling malicious agent interactions with formal verification. *Procedia Computer Science, 95,* 24–31.

Wu, G., Sun, J., & Chen, J. (2016). A survey on the security of cyber-physical systems. *Control Theory and Technology, 14,* 2–10.

Yong, P., Tianbo, L., Jingli, L., Yang, G., Xiaobo, G., & Feng, X. (2013). Cyber-physical system risk assessment. In: *Ninth international conference on intelligent information hiding and multimedia signal processing*, Beijing, China.

Chapter 3
Wireless Sensor Network Security for Cyber-Physical Systems

Wireless Sensor Networks (WSNs) have become a prominent technology for many applications due to its various advantages and feasibility. It has been applied in government, military, transport, health care, education, business, and environment applications. WSN makes sensing, tracking, monitoring, and automation much simpler, effective, and efficient as compared to its predecessor technologies. Due to its vast implementations and immense potential, it has also attracted many threats and vulnerabilities along its path to advancement. This chapter reviews the existing literature and discusses the security issues related to WSN and the challenges in addressing them.

3.1 Introduction

WSN can be described as a network of sensors and actuators connected wirelessly to monitor and control a particular task, operation, or environment (Yang 2014). The basic structure of a WSN, as shown in Fig. 3.1, consists of a sensor field made up of sensors that are placed to gather data about the given task and they report the parameters to a base station or data sink which is connected to the task manager through the Internet or any private network (Khalid et al. 2013; Boukerch et al. 2007; Chen et al. 2007b). Some of the basic advantages of WSN are its low cost, small size, low-power, robustness, ease of deployment, durability, multifunction ability, etc. (Han et al. 2014; Yick et al. 2008; Yang and Cao 2008; Li and Gong 2008). Due to these benefits, WSN finds wide applications in military, transportation, health care, industries, businesses, education, etc., (Han et al. 2014; Yick et al. 2008; Boukerch et al. 2007). Due to their size and toughness, these sensors are also able to be deployed in places like deep underwater, enemy territory, wild forest, extreme hot/cold environment, etc., which was previously impossible for humans to monitor.

© Springer International Publishing AG 2018
S. Ali et al., *Cyber Security for Cyber Physical Systems*, Studies in Computational Intelligence 768, https://doi.org/10.1007/978-3-319-75880-0_3

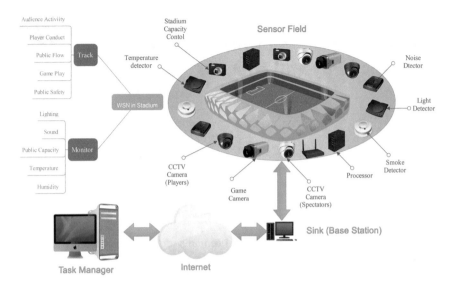

Fig. 3.1 WSN structure

Research in the sensor network technology had started in the late seventies (Lopez et al. 2010). The concept of WSN came into being in the late 1990s and early 2000s, when the technology had matured enough to enable creation of a distributed network of small size, low-power nodes that can communicate wirelessly (Byers and Nasser 2000; Estrin et al. 1999). As research and development of this field advanced further, the IEEE released the standards that came to be known as smart transducer networking (Lee 2000). This specified the desired functions of a sensor node and the technology started to be known as the smart sensors (Lewis 2004). With time, the concept came to be called as WSN. Zigbee standard was later released by IEEE as a wireless communication standard for WSNs (Yang 2014; Egan 2005; Kinney 2003). WSN technology is currently being used in a number of applications as depicted in Fig. 3.2. Just like every other communication technologies, WSN also has its own security goals, issues, vulnerabilities, and threats.

This chapter conducts a survey on the security issues, attacks, and countermeasures implemented in the WSN. This survey also considers the importance of trust and reputation concept in WSN.

3.2 WSN Security

The field of WSN is expanding at an immense pace, and that is attracting large number of researchers, engineers, investors, governments, and most importantly, attackers into this field. Sarma and Kar mention in (Sarma and Kar 2006) that eavesdropping become simple in wireless networks due to its broadcast nature, where wireless medium is inherently less secure. Any transmission can easily be

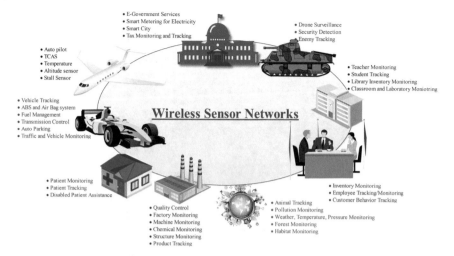

Fig. 3.2 Applications of WSN

interrupted or changed by an opponent. The wireless medium helps an attacker to easily intercept the communication and inject malicious packets. In the absence of satisfactory security, sensor networks deployment is susceptible to number of attacks. Sensor node's restrictions and nature of wireless communication postures exceptional security challenges. As mentioned by Zia and Zomaya (2006), Undercoffer et al. (2002), research in sensor network security is normally conducted in a trusted atmosphere; however, there are number of research challenges still needs to be answered before they are considered trusted sensor networks. These can be clearly stated as of two types namely *Security* and *Reliability*. Reliability is an aspect that ensures that the system performs as per its specifications. Some of the issues concerned with reliability in WSN are identified as detection/sensing, data transmission, data packets, event occurrence (Willig and Karl 2005; Mahmood et al. 2015; Hsu et al. 2007). Security is a major issue in WSN, just like other wireless communication technologies like GSM, Bluetooth, etc. Some of the security issues in WSN are shown in Fig. 3.3 (Yu et al. 2012; Lopez et al. 2009; Li and Gong 2008).

The possible security attacks in WSNs are identified as follows:

- Node subversion
- Passive Information Gathering
- Node Malfunction
- Node Outage
- False Node
- Message Corruption
- Traffic Analysis
- Sinkhole attacks
- Routing loops
- Wormholes

Fig. 3.3 Security issues in WSN

- Selective forwarding
- DoS attacks
- Sybil attacks
- Hello flood attacks

Attacks in WSN can be classified in a variety of ways. From the perspective of the attacker type, it is divided into two types, internal and external attacks (Yu et al. 2012). Internal attacks occur when the attacker gets hold of a node by breaking through the cryptographic processes and external attacks happen when the attacker is able to eavesdrop and inject fractional data without being able to take full control of any part of the WSN. Attacks are also classified based on the layer at which the attack takes place. The layers are physical, link, network, transport and the application layers (Yang 2014). The application layer is the layer with most number of attacks. Some of the attacks suffered by WSNs are classified and represented in Fig. 3.4 (Araujo et al. 2012; Lopez et al. 2010; López and Zhou 2008; Li and Gong 2008).

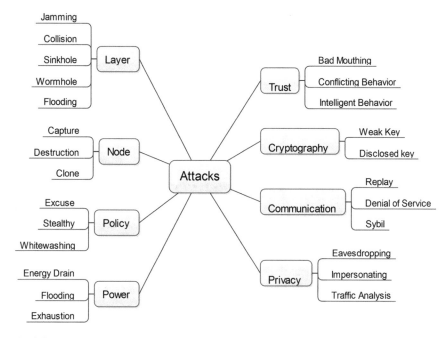

Fig. 3.4 Attacks on WSNs

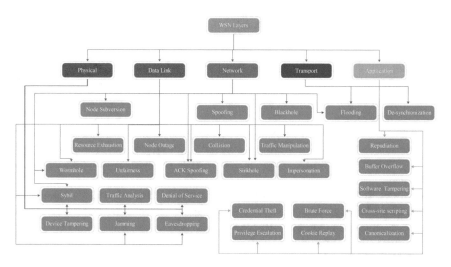

Fig. 3.5 Attacks on various layers of WSNs

Based on the description given in Fig. 3.4, various layers and their attacks have been identified and segmented, while Fig. 3.5 gives a basic understanding of the various layers in a WSN and their respective attacks involved. The layers in the

WSN are divided into five layers named as physical layer, data link layer, network layer, transport layer, and application layer.

The attacks on WSN are of numerous kinds. Many attacks can be implemented on multiple layers. These attacks are grouped according to the layers and discussed in the following sections. The following subsections give a brief discussion on the techniques used to execute the attacks and the effects of successful attacks. The countermeasures or defense implemented to defend against these attacks are also discussed.

3.2.1 Security Issues Related to the Physical Layer

Traffic Analysis (Virmani et al. 2014; Mohammadi and Jadidoleslamy 2011b)

Attacks: Tap network information patterns which may result in deteriorated network performance, high packet collision, traffic alteration.

Technique: Network data flow monitor examines packets by packet.

Defense: Critical observation of unfair collision, misbehavior, and identity, standard link-layer encryption, MAC request rate, and use of large packets.

Eavesdropping (Virmani et al. 2014; Mohammadi and Jadidoleslamy 2011b)

Attack: Extract vital WSN data.

Technique: Pay close vigilance on the communication channel by collapsing wireless network transmission medium.

Defense: Protection involves solutions like systematic access control, dispersed processing, access constraint, advanced encryption, and peripatetic security.

Jamming (Sun et al. 2014; Sabeel et al. 2013; Mohammadi and Jadidoleslamy 2011b)

Attack: Interference is introduced intensively to disrupt WSN radio signals of identical frequencies.

Technique: Constant jamming, deceptive jamming, random jamming, and reactive jamming to cause energy exhaustion, disruptive communication, block entire bandwidth, corrupt data packets, and deceive network's defensive mechanisms.

Defense: Detected using statistical information, channel utility degradation threshold, and background noise. The WSN can be protected by employing access restriction, strong encryption, CRC check, low duty cycle, high broadcast power, hybrid Frequency hopping spread spectrum/direct sequence spread spectrum (FHSS/DSSS), ultrawideband, antenna polarization, directional transmission, and jamming avoidance itinerary design (JAID).

Sybil Attack (Virmani et al. 2014; Sabeel et al. 2013; Mohammadi and Jadidoleslamy 2011b)

Attack: Cause network inaccessibility.

Technique: Network sabotage by counterfeiting characters by manifold fake nodal identities.

Defense: Detection includes the use of low overhead and signal delay. A physical shielding of nodes is necessary to defend against Sybil attack.

Path-Based DoS (Jadidoleslamy 2014; Mohammadi and Jadidoleslamy 2011b)
  Attack: Exhaust nodes' battery, disturb network, and reject node.

  Technique: Broadcast of large data packets to create mixed jamming attacks.

  Defense: Enhancing redundancies, anti-attack capability, acknowledge validation, and gray-listing.

3.2.2 Security Issues Related to the Data Link Layer

Node Outage (Jadidoleslamy 2014; Mohammadi and Jadidoleslamy 2011a)
  Attack: Gain control of the functionality of WSN's components, tap critical information, kill legitimate nodes to halt its services, and inserts malicious data.

  Technique: Capture and reprogram normal nodes and electronic device. Attacker cessation of network nodes, consistent node observation,

  Defense: Internodal support, and operation are key aspects of its discovery. Provision of an alternative path, tough protocols, defending physical, and node capture attacks are its countermeasures.

Jamming (Jadidoleslamy 2014; Mohammadi and Jadidoleslamy 2011a)
  Attack: Create intensive packet collision and nodal resource exhaustion.

  Technique: Jamming of sensor-medium access control (S-MAC), Berkeley media access control (B-MAC), and lightweight-medium access control (L-MAC) protocol packets.

  Defense: Detection includes low false alarm rate, proactive detection, and quick detection. Countermeasures include limiting MAC request rate, using small sizes, defending S-MAC, protocol mapping, and employing wormhole-based anti-jamming Techniques.

Collision (Sun et al. 2014; Jadidoleslamy 2014; Mohammadi and Jadidoleslamy 2011a)
  Attack: Create an environmental collision, probabilistic collision, and ACK alteration. Interferences corrupt/discard data or control packets and exhaust the energy.

  Technique: Simultaneous broadcast of the message of the same frequency by two nodes.

  Defense: Enabling time.

Resource Exhaustion (Sun et al. 2014; Jadidoleslamy 2014; Sabeel et al. 2013; Mohammadi and Jadidoleslamy 2011a)
  Attack: Destruction of the sensor.

Technique: Repeated collisions and retransmission, attack on request to send/clear to send (RTS/CTS) and later in acknowledgment (ACK).

Defense: Detected by finding misbehavior. Countermeasures include MAC rate restriction, random backoff rate, TDM, regulated link response rate, and ID protection.

Traffic Manipulation (Virmani et al. 2014; Sabeel et al. 2013; Mohammadi and Jadidoleslamy 2011a)

Attack: It creates aggressive channel usage, ineffectual network, traffic distortion, increases contention, deteriorated signal quality.

Technique: Build constraints for infected MAC protocol and imitate work just like a normal node by monitoring channel and floating MAC scheme usage.

Defense: Its countermeasures include traffic analysis, collision defenses, misbehavior identity, link-layer encryption, and regulated MAC requests.

Unfairness (Jadidoleslamy 2014; Mohammadi and Jadidoleslamy 2011a)

Attack: It reduces efficiency and channel access capacity.

Technique: Partial DoS attack with collisions and resource exhaustion and misuse of MAC-layer priority.

Defense: Detection of MAC-layer discrepancies.

Acknowledgment Spoofing (Sabeel et al. 2013; Mohammadi and Jadidoleslamy 2011a)

Attack: It causes loss of packets, steers loops, broadcast blunder messages.

Technique: Send-up of link-layer nodal ACKs of overheard packets, modify/replay tracking data.

Defense: Countermeasures include the use of the new route, validation, link-layer encryption, and global-shared key techniques.

Sinkhole (Virmani et al. 2014; Jadidoleslamy 2014; Mohammadi and Jadidoleslamy 2011a)

Attack: Stuck traffic, eavesdrop on selective forwarding, black hole and wormhole attacks, annex base station, modification of messages, and route table.

Technique: Resources exhaustion, advertise wrong routing information to forward packets to the hub, trap nodes, compel application data packets, fabricate information, identify spoofing.

Defense: Hop-count monitoring scheme, monitoring node's CPU usage, and using message digest algorithm.

Eavesdropping (Sabeel et al. 2013; Mohammadi and Jadidoleslamy 2011a)

Attack: Exploit transmission medium, extract critical data, and expose privacy.

Technique: Interception in network.

Defense: Countermeasures include control of access, distributed processing with strong encryption.

Impersonation (Jadidoleslamy 2014; Mohammadi and Jadidoleslamy 2011a)

Attack: Deactivate cluster leader to divert nodes.

Technique: Attacker reidentifies dummy MAC addresses, gains physical access, modifies routing table, hypnotizes nodes to disrupt routing tables, kills sensor, congest and split network, generate deceitful data resources, leak encryption keys, and critical information.

Defense: Detection includes detection of false identity, misconduct, deceitful routing, and detect collision. Countermeasures include strong authentication, secure routing, proof techniques, secure identity, limited MAC rate, small packet frames.

Wormholes (Virmani et al. 2014; Jadidoleslamy 2014; Mohammadi and Jadidoleslamy 2011a)

Attack: To create false routes, chaos in WSN, overused routing race conditions, change in network topology, the collapse of path detection protocol, packet destruction.

Technique: Transmit information secretly between two WSN nodes targeted with false and forged routing information.

Defense: Detection of false routing information helps detect wormhole attacks. Countermeasures include multidimensional scaling algorithm, DAWWSEN protocol, border control protocol, graphical position system, ultrasound, sync in global clock, and authenticated encryption of link layer.

Sybil Attack (Virmani et al. 2014; Sabeel et al. 2013; Mohammadi and Jadidoleslamy 2011a)

Attack: Cause network ineffectiveness in data integrity and accessibility.

Technique: A reputation system is sabotaged by counterfeiting characters in peer-to-peer networks. Attacker node manifolds fake nodal identities.

Defense: Detected by maintaining a low overhead and signal delay. A countermeasure is the physical shield of nodes with regular change of key.

3.2.3 Security Issues Related to the Network Layer

Wormhole (Virmani et al. 2014; Singh et al. 2014; Kaur and Singh 2014; Jadidoleslamy 2014)

Attack: Results in false routes, chaos in WSN, overused routing race conditions, change in network topology, the collapse of path detection protocol, and packet destruction.

Technique: Transmit information secretly between two WSN node targeted with false and forged routing information.

Defense: Discovery includes detection of false routing information, detect wormhole attacks. Countermeasures include multidimensional scaling algorithm, DAWWSEN protocol, border control protocol, graphical position system, ultrasound, sync in the global clock, authenticated encryption of link-layer and global-shared key.

Sybil Attack (Virmani et al. 2014; Kaur and Singh 2014; Yu et al. 2012)

Attack: Causes network ineffectiveness in data integrity and accessibility.

Technique: A reputation system is sabotaged by counterfeiting characters in peer-to-peer networks. Attacker node manifolds fake nodal identities.

Defense: Detection method is maintaining a low overhead and signal delay. Countermeasures are the physical shield of nodes with regular change of key.

Sinkhole (Virmani et al. 2014; Singh et al. 2014; Kaur and Singh 2014; Jadidoleslamy 2014)

Attack: Its effects include traffic suck up, eavesdropping, selective forwarding, black hole and wormhole attacks, annexure of base station's position, information and packet alter, selective messages suppression, route table modification, and exhaustion of resources.

Technique: The attacker advertises wrong routing information. All data network packets are drawn to a centralized compromised hub close to the base station to enable selective forwarding to launch secondary attacks. It traps and hypnotizes nodes, compels application data packets along the flow path, alters receiving traffic information.

Defense: Detection includes false routing information of dynamic neighboring nodes, verification by tree nodal structure visual geographical map. Countermeasures include hybrid intrusion detection system (IDS), sensor network-automated intrusion detection system, detection on mini route, probabilistic next hop count, authentication, link-layer encryption, global-shared key techniques, routing through access restriction, and detection of wormhole attacks.

Black Hole (Virmani et al. 2014; Sun et al. 2014; Singh et al. 2014; Yu et al. 2012)

Attack: Its suppresses broadcast, attracts all network traffic into the false shortest path thus creating a black hole, disrupts network routing table, decreases network throughput, network is split, and increases packet loss rate.

Technique: Suppression broadcast messaging to affect overall traffic flow. In a black hole attack, malicious nodes do not send true control messages. A false route replies (RREP) message from the adjacent node is sent in place of it with the shortest path. Hence, message transfer takes place to the attacker.

Defense: Detection methods are abnormal time difference route request (RREQ) and the required number of packet transmissions (RNPS). Countermeasures are authenticated surveillance and redundancy, multipath routing, decentralized intrusion detection system, and sensor network-automated intrusion detection system.

Spoofing (Singh et al. 2014; Yu et al. 2012)

Attack: The results are network split, overrun of resources, lifetime reduction in network, shedding of routing data. It leads to threats like integrity and authenticity.

Technique: To take off added network devices to launch attacks against other network hosts. The attack is done by creating a loop between the source and destination nodes.

Defense: It can be discovered by implementing secure ARP protocol, kernel-based patches, and passively static MAC. Some of the protective measures are encryption with MAC using a different path for resending messages.

Acknowledgment Spoofing (Singh et al. 2014; Jadidoleslamy 2014)
Attack: It causes loss of packets, steers loops, broadcast blunder messages. Modify and replay tracking data.
Technique: Send-up of link-layer nodal ACKs of overheard packets.
Defense: Countermeasures include the use of the new route, validation, link-layer encryption, and global-shared key techniques.

Flooding (Virmani et al. 2014; Singh et al. 2014; Yu et al. 2012)
Attack: Exhaustion of resources, retarded availability, and downgraded traffic.
Technique: Producing and proliferating plentiful useless route requests. Various types are broadcast flooding, target flooding, false identity broadcast flooding, and false identity target flooding.
Defense: Bidirectional authentication.

3.2.4 Security Issues Related to the Transport Layer

De-synchronization (Sun et al. 2014)
Attack: Creating a disruptive network and resource collapse.
Technique: Disrupting re-synchronized connections between nodal transmissions.
Defense: Detection methods are delay and warping in performance. Countermeasures are dual path check.

Flooding (Virmani et al. 2014; Sun et al. 2014; Jadidoleslamy 2014)
Attack: Effects are exhaustion of resources, retarded availability, and downgraded traffic.
Technique: Producing and proliferating plentiful useless route requests. Various types are broadcast flooding, target flooding, false identity broadcast flooding, and false identity target flooding.
Defense: Bidirectional authentication.

3.2.5 Security Issues Related to the Application Layer

Repudiation (Sun et al. 2014)
Attack: Absence of controls to track proper actions of users, facilitating malicious manipulations of identity and data.
Technique: Launch of the selective forward attack.

Defense: Detection includes false, deceptive log files. Countermeasure includes sensor node identity and detection.

- **Buffer Overflow** (Cowan et al. 2000; Silberman and Johnson 2004)

Attack: Effects include the destruction of program action, memory inaccessibility, unfitting results, and memory thrashing.

Technique: Writing data exceeding the buffer and memory capacity in a program. Input results are exceeding abnormally.

Defense: Detection includes modifications in static analysis, compiler, and operating system. Countermeasures include strong program language, safe libraries, protection of buffer overflow, pointers, executable space, random address space layout, and inspection of the deep packet.

- **Cross-site Scripting** (Patil et al. 2011; Choi et al. 2012)

Attack: Avail access and edit cookies, session tokens, and Hypertext Markup Language (HTML) page.

Technique: Inject dangerous scripts into user's websites.

Defense: Detection includes changes in Uniform Resource Locator (URL) scripts. Countermeasures include context-sensitive server encryption and benign JavaScript APIs.

- **Canonicalization** (Lynch 1999)

Attack/Technique: Replicating data into various forms with malformed representations.

Defense: Detection includes altered Hypertext Transfer Protocol (HTTP) Posts, HTTP Gets, NET: Unicode, JSP, Java: Unicode, Personal Home Page (PHP). Countermeasures include robust encoded internationalized Unicode input.

- **Software Tempering** (Xing et al. 2010; Sastry et al. 2013)

Attacks/Technique: Modification of application's runtime in effect behavior to carry out unlawful attacks resulting in misused binary patch up and code replacement.

Defense: Countermeasures include anti-tamper software, malware scanners, and antivirus applications.

- **Brute Force Attack** (Benenson et al. 2008; Becher et al. 2006; Fatema and Brad 2014)

Attack/Technique: Huge sample space of key is systematically searched for cryptanalysis.

Defense: Countermeasures include limiting a number of attempts and securing accounts after unsuccessful logins.

- **Cookie Replay** (Liu et al. 2005)

  Attack: Cause network masquerade.
  Technique: Nasty repeated and delayed valid data transmission by data diversion and seize by substitute IP packet.
  Defense: Detection includes data conflict caused with replay and message travel via untrusted mediators.

- **Credential Theft** (Baig et al. 2012)

  Attack: Capture account credentials of a computer to validate other computers in the network.
  Technique: Lateral movement and privilege escalation.
  Defense: Detection includes abnormal functions of user accounts. Countermeasures include restriction of confidential domain and local accounts, inbound traffic, users from the local administrator's group, privileged domain accounts, and outbound proxies' remote management tools.

- **Privilege Escalation** (Amini et al. 2007)

  Attack: Unlawfully gain access to protected resources for unauthorized actions.
  Technique: Look out for weak bug, design flaw, or configuration in the operating system.
  Defense: Detection includes abnormal user access and unusual changes in the network. Countermeasures include protection of data execution, random address space layout, least privilege applications, a digital sign of kernel mode, antivirus software, patching, compilers to trap overrun of the buffer, encryption, and mandatory access controls.

  The next section introduces the concepts of trust and reputation in WSN. A thorough literature review of its advances in WSN has been conducted and how it can be linked with security of WSN to reduce the security-related threats and its vital nature has also been discussed.

3.3 Trust and Reputation—Their Importance in WSNs and to Improve Its Resilience

The concept of trust and reputation exists since the dawn of civilization and has been studied by different researchers in different fields (Hardin 2002; Wright 2010; Ashraf et al. 2006). Observing this behavior in human beings, researchers have tried to embed it into the context of information technology. The concept of trust and reputation in the field of WSN was initiated during the mid-2000s and has been advancing since then.

The previous section provides an in-depth study and analysis of the various kinds of security issues and attacks suffered by WSN and some of their counter-measures being implemented in the current times. Although these security techniques have helped the WSN advancement to a much secure state, but the technique of trust and reputation forms one of the most effective methods of obtaining better security in the WSN. Trust is defined as the degree of belief about the behavior of a given node based on the past experiences with it. Reputation is defined as the global perception of a particular node and it is based on the trust that third-party nodes holds on it (Boukerch et al. 2007). Trust eases various other processor/ resource-hungry and complex security techniques like authentication, cryptography, etc., since they can be made hard or soft based on the trust a node has on the other node. It not only helps achieve a secure relationship among the nodes, but also helps the nodes in decision-making and promotes collaboration. As pointed out by Yu et al. (2010), trust is context-sensitive, subjective, unidirectional, and transitive. It is a vital security measure in this case because it is an inexpensive and effective solution for the low-capacity devices implemented in WSNs.

The basic parameters of ensuring trust are the trust qualification and the trust computation (Pirzada and McDonald 2004). Trust qualifications represent the various levels of trust, while the trust computation defines the method of computing the trust value among nodes. The scheme, method, or procedure that defines the various steps and parameters that govern trust and reputation is termed as Trust and Reputation Management scheme (TRM). The term Trust Management was introduced by Blaze et al. (1996). Initially, at the infancy stage of WSN, the paradigm was mostly connected to an ad hoc mobile network (Pirzada and McDonald 2004; Liu et al. 2004), P2P network (Kamvar et al. 2003) or MANETs (Jiang and Baras 2004). The basic steps in a TRM are to collect the trust data, update the trust values, and to execute the decision-making.

3.3.1 Surveys on Trust and Reputation

It is well understood from the discussion in Sect. 3.1 and the illustration in Fig. 3.1, that one of the vital properties of the WSN is its distributed nature. This has been highlighted in the survey conducted by Sorniotti et al. (2007) which illustrates it by giving importance to the distributed processing of the data within the network. It is due to the fact that it makes the communication and data transfer much simpler. It points out many advantages of the technique and the trust involved in it. The authors discuss the sensor node failure detection schemes like self-diagnosis and the group detection methods. The reputation systems and a trust-based framework have also been studied and discussed. Reputation system by (Ganeriwal et al. 2008) forms a blend of domains like economics, statistics, data analysis, and cryptography and this helps establish the trust within the sensor nodes. Another approach discussed was the Bayesian approach (Ganeriwal et al. 2008) in which the node maintains its trust and reputation values on the other concerned nodes. Trust-based frameworks for noncritical sensor networks using cryptographic material (Anderson et al. 2004), public-key authentication (Ngai and Lyu 2004) and Bayesian and beta distribution probability (Zhang et al. 2006a) have been discussed. The next notable survey was by Yick et al. (2008). It provides a broad picture of WSN by giving a good literature review on its various issues and standards. Localization, synchronization, and coverage of the sensors and their security are some of the major issues discussed. Distributed reputation-based beacon trust system (DRBTS) (Srinivasan et al. 2006) is used for the secure localization of nodes. The SecRout protocol (Yin and Madria 2006) is used for packet delivery. The secure cell relay (SCR) protocol (Du et al. 2006) is used for attack resistance. All of the three approaches above use the trust system.

Srinivasan et al. (2009) provided a thorough survey focusing on trust and reputation systems. They discuss the social perspective like trust, belief, uncertainty, etc., in detail and the network perceptive like mobile ad hoc network (MANET) (Govindan and Mohapatra 2012), WSN, node misbehavior. Reputation and trust-based systems have been reviewed in aspects of its goals, properties with fitting initializations, pros, cons and cataloging with an in-depth citation. Lopez et al. (2010), conducted an extensive and exclusive survey on the trust management systems in WSNs. Since the field of WSN was well established by then, they were able to present a list of best practices based on their study and understanding. The basic steps for any trust and reputation system are to set up the nodes, initialization of the trust parameters, information gathering, parameter updating, and the evaluation of risk. The steps have been further simplified and broken by them and explained in detail. An advancement of their survey can be found in the survey presented by Khalid et al. (2013) in 2013. They conducted an extensive survey which concentrates exclusively on the trust and reputation for WSNs. The cryptographic technique like a symmetric key with authentication using hash functions and implemented digital signatures were discussed. The concepts of trust and reputation observed in human behavior have been implemented into the WSN for node interactions. The various properties of trust are subjective, asymmetric, reflexive, partial transitive, and content-sensitive nature. The concept of trust has been studied and discussed in various domains like p2p networks, grid computing, opportunistic computing, social networking, e-commerce, and the WSNs.

Han et al. (2014) presented an individualized survey of the nature and applications of trust models for WSN for detecting unusual nodal behaviors. The importance of trustworthiness needs to establish trust metrics and preserve privacy at the same time. Future studies state their enhanced use in other WSN types. Some of the other eminent surveys are Srivastava and Johri (2012) and Reshmi and Sajitha (2014). They have individually presented an expounded review of various trust management schemes which employ indirect and direct methods to compute trust and identify its main features, pros, and cons and open issues for future research. Both papers stated that the presence of both QoS and social trust parameters are essential for trust.

3.3.2 Trust and Reputation Management (TRM)

TRM is the procedure of managing the trust of the nodes and the components in a WSN. The current section discussed the various TRM procedures proposed and implemented in WSN over the years.

One of the earliest TRM related to the WSN was proposed by Pirzada and McDonald (2004). The paper proposed a trust model which consists of three functions, trust deviation, qualification, and computation. The deviation was basically the reputation of the node obtained from other nodes in the network based on eight different parameters. The qualification scale is in the range from −1 to +1. For trust computation, a relationship was given which involves the situational trust and the weight. This brought about the interest of other researchers around the world into this field. One such work was done by Momani et al. (2007a, b). Momani et al. took the concept of beta reputation system used in e-commerce in (Jsang and Ismail 2002) and implement it in WSN. Momani et al. (2007b) conducts a survey and proposes a method to model trust in WSN. This was further developed to propose the recursive Bayesian approach (Momani et al. 2007a). Momani and Challa (2008) argue that a malicious node may still be trusted if it conducts seamless communications while injecting false data into the system. Thus, the data trust was introduced with the existing communication trust to form an improved trust model. Then in Momani et al. (2008), they compare their model with another model to prove that considering just the communication trust is insufficient to trust a node.

Chen et al. (2007a), exemplified the lack of security being implemented in WSN thus proposing a trust management framework incorporating the tools from statistics, probability, and mathematical analysis. Thus, it uses a mathematical framework for trust and reputation. Chen et al. (2007a) also introduced the term certainty, which defines the amount of certainty in the values of trust and reputation of a given node. It states the outcomes of positive or negative in deficient for an adequate decision-making of a scenario incurred in WSN. To build on it, they have projected at a four-point formula of firstly bestowing a reputation and trust space which are interconvertible with a clear explanation in this regard. A watchdog mechanism is used which takes cares of the three parameters, namely trust, reputation and certainty of each node in the WSN. Similar to this, another group of researchers, Fernandez-Gago et al. (2007), yet again emphasizes the lack of proper trust management solutions in WSN when compared to ad hoc and P2P networks. The hitches with its potential solutions in the existing system are considered to be further enhanced for its usage in WSN. The prominence of a history-aware system and the role of the base station were evidently explained (Fernandez-Gago et al. 2007).

It is understood from the discussions in the introductory sections that the sensors in the WSN works as a group to monitor and track any object. Also, the TRM depends heavily on the grouping nature of WSN. Shaikh et al. (2009), in 2009, took this into consideration and proposed a lightweight group-based trust management scheme (GTMS) for clustered WSN to provide faster trust evaluation. Results with faster memory employing less energy consumption and less overhead for communication with the suitability for large-scale WSN had been obtained. GTMS is also equipped with the enhanced feature of detecting abnormal nodes at the earliest.

To help the nodes to reduce the burden of computing and updating the trust and reputation parameters, Boukerch et al. (2007) proposed an agent-based scheme in 2007. It was one of the earliest research which started using the term WSN and implementing trust and reputation into it. They proposed the agent-based trust and reputation management scheme (ATRM) based on clustered WSN whose core is a mobile agent system. Each node consists of a t-instrument that stores the trust value and a r-certificate that stores the reputation value of the node and a mobile agent is assigned to administer these values. This tries to tackle the network performance issue which was neglected by the earlier researchers. By this, the authors reduce the overhead in terms of time delay and extra packet transmission which was proved in the simulation conducted by them. Reddy and Selmic (2011) developed on this method and proposed an agent-based trust calculation scheme in 2011. It eliminates the requirement of having a high-end resource for trust computations. This technique was specifically developed for WSNs taking care of the limited resources for storage, computation, and communication. It uses an agent that collects the reputation of each node within communicating distance in a cluster. This agent calculates the trust of each node based on the information gained. The authors propose two different ways of calculating the trust based on the reputation of the node in its particular cluster, using the collaborative approach. This lowers the computational work of the node and thus increases its durability.

Since the concept of TRM and the field of WSN are in its early stage of development, the standard in this field is yet to be defined (Mármol and Pérez 2010). This made the comparison among TRMs a difficult task. Marmol et al. (Mármol and Pérez 2009) in 2009 have tried to solve this problem by building a simulator. They stated the fact of the incapability to check the correctness of the model in order to compare with another trust model. They have proposed a Java-based trust and reputation model simulator TRMSim-WSN in order to provide the trust of the system in an easy-going manner. It is a user-friendly system with various provisions to adjust parameters related to nodes or the possibility of collision in order to carry out customized simulations.

Hierarchical trust management is one of the techniques for TRM. Bao et al. (2011a, b, 2012) have done extensive research in this field. They proposed an extremely scalable cluster-based hierarchical trust management protocol in 2012. It was specifically designed for WSN networks. It improved the earlier researchers' work by creating multidimensional trust attributes for overall trust determination. The comparison of subjective trust resulted from protocol execution time and objective trust from actual node status for validating their protocol design is an innovative concept. Their application for routing and intrusion detection has approached ideal performance in flood-based routing and optimal trust threshold respectively. Their trust-based IDS algorithm beats traditional methods. Their future scope includes providing decentralized trust management and hierarchical trust management for dynamic WSN.

3.3.3 Trust Models and Frameworks

Trust model gives a complete and comprehensive description of the trust and reputation implementation and management in the WSN system. Researchers have proposed models based on concepts that include statistical methods, fuzzy logic, analytical method, nature-inspired method, etc. The orientations of WSN like distributed, individual, clustered etc. were considered. This section presents a study of the advances on various trust models through the years.

Clustering is a method in which a number of similar nodes are grouped together to form a cluster. A cluster head is appointed by the group, which takes care of the communication between the cluster and the base station. Based on this concept, Srinivasan et al. (2006) proposed the distributed reputation-based beacon trust system. Beacons are the nodes that help the sensor nodes to derive their location, and DRBTS helps identify the malicious nodes based on its reputation and a majority voting scheme. Probst and Kasera in (2007) took a statistical approach toward the trust establishment in WSNs. The direct and the indirect experiences of the nodes are used to evaluate the confidence interval and this helps understand the tradeoff between tightness of trust and the resources being used. Dynamic scaling of redundancy is used to reduce the power consumption. Crosby and Pissinou (2007) proposed the cluster-based reputation and trust model. The paper pointed out the benefits of energy efficiency, easier data aggregation, and suitability for high-scale networks, of this approach. A valuable research had been conducted to understand the consequences and effects of having a bad node within a cluster or system. The model helps good nodes to build on their trust value over time and vice versa. A probabilistic approach is used to build the model. The simulations by the authors had shown feasible results. This was further developed by Shaikh et al. in (2009).

Localization is yet another issue in the WSN. It is very critical for the sensor nodes and the base station to know the exact locations of the nodes. This helps to understand what and where exactly are the data which is being sensed. One of the earliest papers on this was by Zhang et al. in (2006b) which proposed a location-based authentication using ID-based cryptography (IBC). Although the proposal was not exclusive for WSN, it did provide a beginning to the research on localization. Boukerche et al. (2008) studied various security methods in WSN and considered them for implementation for localization. Crosby et al. (2011) realized the importance of localization and proposed a trust-based detection with location-aware capability. This was a development of their previous work in (Crosby and Pissinou 2007). The location awareness helps their model to reinforce the integrity. It was done by the simple technique of analyzing the strength of the received signals and verifying the location. Based on the threshold for the difference between the reported location and the computed location, the decision of the nodes authenticity is made. A dynamic updating scheme is used which periodically verifies the locations of their one-hop neighbors. The location parameters are updated along with the trust and reputation parameters. One of the current researches is by Miao et al. in (2015), which proposes a lightweight distributed localization scheme. A three-step localization technique termed as Virtual-force Localization Algorithm (VLA) is proposed and implemented. The location is then verified using the received signal strength indicator (RSSI) and a decision on the good node, drifting node, or malicious node is made.

Kim and Seo (2008) introduced the concept of fuzzy logic into the WSN trust. It helps to distinguish between the good and the bad sensor based on the trustworthiness and trustworthiness values of the sensor. These values are implemented on the fuzzy logic relationships and the corresponding result gives the level of trust on a particular node. Though it brought a new idea into the field, it was untested for its feasibility. A different idea was presented by Zhan et al. in (2009) by introducing the SensorTrust model. It is a model initiated for hierarchical WSNs to mitigate problems mainly concerned with data integrity and it can incorporate robustness as well. In this model, trust is estimated for children nodes as well. It is a dynamic memory-based system integrating the past risk history to the current values. The value helps identify the current trust scenario by employing the Gaussian model (Fraley 1998). The model is used to rate the integrity of data with the help of protocol and to adjust to diverse perspective. Results show that SensorTrust promises to identify the level of trust in WSN against attempts to report false data.

Naseer (2012), explicitly stated the essential provision of security against behavioral-related attacks, in particular, trust-aware routing for varying types of WSN circumstances. A trust-based routing framework had been framed with various reputation systems explaining an in-depth examination of its pros and cons. Also, an innovative concept of sensor node-attached reputation evaluator (SNARE) for accomplishing the needs of WSN had been framed to monitor, rate and responded its components to apply optimized condition for WSN. Geographic, energy and trust-aware routing (GE-TAR) promised an increased packet flow rate during the process.

As the WSN and trust became and established the field, energy efficiency became an important aspect of the WSN. The work by Almasri et al. (2013) is one such research addressing this issue. TERP (Almasri et al. 2013), is a routing protocol that takes care of the trust and the energy efficiency in the WSN. It is based on the destination-sequenced distance vector (DSDV) protocol.

One of the latest papers in the area was by Karthik et al. (Karthik and Karthik 2014) in 2014. They exemplified the critical value of trust to bring about reliability for WSN. Thus, a framework that evaluates trust mechanism in contemplation to the underlying system challenges was proposed. The model is based on three major factors, security, reliability, and mobility, which is an advancement to the previous work in Dhulipala et al. (2013) in real time. In future, more dimensions such as accuracy, scalability, and fault-tolerance are planned to be included.

The latest research on this is to create a lightweight model which makes the trust model easier, simpler and faster. For example, Li et al. (2013) proposed a lightweight trust system for WSN known as Lightweight and Dependable Trust System (LDTS). The energy consumption is reduced by eliminating the feedback of the cluster nodes and by implementing dependability enhanced trust mechanism. A self-adaptive weighted method is used for trust aggregation. These made LDTS more efficient as it required lower memory and power. Wang et al. (2014) put an effort to make the computations simpler and energy consumption lower. Either of the two kinds of trust, direct or indirect, can be computed from the trust matrix, based on the understanding between the nodes. Based on the algorithm, the decision is made. The model is a tradeoff between the detection rate and the energy consumption. The research by Singh et al. in (2015) develops this further by proposing the lightweight trust model (LTM). The model has a dynamic trust-building mechanism which is also embedded with priority task option. The trust aggregation is executed using a self-adaptive weighted method and helps reduce the misjudgments. Che et al. (2015) incorporate Bayesian (FranzÈn 2008) and Entropy (Shannon 1948) in Cluster approach. This model implements the Bayesian method to evaluate the trust. A decay factor is involved for the trust updating. It is done by taking care of the node history and the decay factor. A confidence factor is also involved to validate the trust value. Entropy theory is used for this purpose, which distributes the weights to the trust values. This approach makes the model efficient and lightweight since the history gets decayed reducing the memory requirement.

3.3.4 Challenges and Open Research Issues

The two most concerning issues of WSN are the security and reliability. The major challenges to the security of WSN include privacy, environment complications, protected collection, trust management, and topology complications (Anand et al. 2006). Reliability issue comprises of detection/sensing, data transmission, data packets, event occurrence (Willig and Karl 2005; Mahmood et al. 2015; Hsu et al. 2007). The challenges and issues in security and reliability are discussed in this section.

3.3.4.1 Security Challenges

- **Assessing Privacy** (Kumar et al. 2014; Anand et al. 2006): A sensor network being a collection of several different sensing and communication elements, possesses an inherent risk of being individually subject to physical threat which could jeopardize the privacy of the data it holds. Although the compromise of a single sensing node does not result in absolute loss of security of the network, the system would still have to bear the losses incurred. The challenge would be to determine metrics based on logical evidence to deliver a probabilistic guarantee about the level of compromise should there be an attack.

- **Environment Complication** (Huang et al. 2010; Anand et al. 2006): For a sensed value to be meaningful, it is desirable that the location and time parameters of the context also be relayed along with the readings. This gives rise to two concerns. The first one being, the exposure of context and metadata increases the possibility of the attacker deducing patterns among adjacent readings. Identifying a cost-effective arrangement for nondisclosure of the recordings parameters would be the second challenge. These issues could possibly be addressed by introducing breaks between scheduled message relays, or by formulating an algorithm to add a random delay between transmissions or by forging the message sequence to mask the legitimate context information.
- **Protected Collection** (Cardenas et al. 2008; Anand et al. 2006): A conflict of interest has been observed in the standard security policy as opposed to the procedure in practice. The policy specifies the encryption of all traffic that is communicated and that the message should be decrypted only by the source and destination nodes. This highlights that the communication network should not be trusted. On the contrary, since each of the nodes through which the data is passed, should analyze the data, where end-to-end encryption from the sensors to the base station is not possible. This poses a challenge to develop techniques that would not compromise security while allowing the aggregation of data in nodes.
- **Ad hoc networking topology** (Sarma and Kar 2006, 2008): The WSN mostly uses the ad hoc networking topology. The topology is susceptible to various kinds of spoofing attacks and link outbreaks. The attacks can be performed on any node of the WSN based on the attacker's preference. The attack can facilitate leakage of sensitive/secret data/information, interference of messages and imitation of nodes. The imitation attack is easy because of the dynamic nature of WSN, where each node can be activated or deactivated based on the requirement in real time. Node failures are common in WSN due to which the dynamic approach is adopted in WSN. Also, WSN consists of a large number of sensors. The deployment of the sensor is completely based on the location/ object to be monitored. The distribution of the sensor varies according to the desired sensor concentration for the given event to be monitored. Thus, the deployment of WSN sensors is completely irregular. This creates the vulnerability to attacks.
- **Topology Complication** (Anand et al. 2006; Blilat et al. 2012): The topology of WSN is irregular. It is obvious that the data contained by the middle nodes is greater than that by the end nodes. Due to this the issue of data aggregation and irregular distribution occurs in the WSN. Attacking a leaf/end node yields a minimal value to the attacker as its effects of disruption and eavesdropping are small. Thus, attacking a root/middle node is highly beneficial to the attacker as it can gain significant influence on the network. The attack leads to snooping, where the malicious node gets hold of the private node data based on the compensated aggregation performed by the other nodes to balance the

compromised node. Thus, the challenge is to hide the routing structure and table of the network. If the attacker can decipher the routing structure, an attack can be executed by taking hold of high-value locations of the routing tree. This is very advantageous to the attacker since a very few nodes are attacked to capture the complete network.

- **Accessible Trust Management** (Momani et al. 2008, 2015b; Momani and Challa 2010): Trust management is established in WSN to maintain a tolerable level of trust among the node in the network. The most vital function is to clearly distinguish between the genuine and fake nodes. Energy, memory and processing constraints of the WSN nodes are the issues expected to be addressed by trust management approach. The building of trust in the newly deployed nodes and previously attacked nodes is another challenge faced by trust management. The judgment on the number of nodes responses to be considered for a decision-making process is also a complicated task. It is desired to have a robust and lightweight key management system which could be suitable for the large-scale network of WSN.
- **Wireless Media** (Rehana 2009; Wang et al. 2006b): WSN is required to be portable and low cost. Due to this, the wireless media communication is a key aspect of WSN. The media used are radio signals like Bluetooth and Zigbee. The security schemes designed for wire-based communication become invalid in the field of WSN. Thus, it is a challenging task to devise security schemes that fit with the wireless characteristic of WSN.

3.3.4.2 Reliability Challenges

In WSN, the dispersed nodes in different geographical location forwarded all the collected data to the base station (sink). Due to error-prone nature of wireless links, in WSN, reliably transferring data from one node to sink is a major task. Mahmood et al. in (2015) pointed out that reliability of WSNs can be classified into different levels namely packet or event reliability level or hop-by-hop reliability. The packet reliability requires that all the collected packet data from sensor nodes transfer wirelessly to the sink. On the other hand, in event reliability, instead of sending all the information, only certain events have to be sent to the sink. In hop by hop, the intermediate nodes are responsible to perform loss detection and recovery. Whereas, in end-to-end reliability, only the end nodes are responsible to perform error recovery operations (i.e., only the source and destination nodes) (Mahmood et al. 2015).

Few critical WSN applications (e.g., combat zone surveillance applications, intrusion detection applications, etc.) require high or even total end-to-end reliability. Few of them require packet-driven consistency while others only require event-driven reliability (Pereira et al. 2007). Wang et al. (2006a) and Cinque et al. (2006) raised a question "whether hop-by-hop reliability at the transport layer can replace (and even be more efficient than) link-layer reliability at the MAC layer"

(Wang et al. 2005), without ignoring the fact that a minimum reliability grade is required by most routing protocols to attain acceptable degrees of efficiency.

The reliability of data transmission in WSN is threatened by several faults. The power constraints in WSN are forced to use the less power consuming error control techniques, which may lead to the packet lost or delivery of data with errors (Cinque et al. 2006).

Abouei et al. (2011) enhanced reliability in transmission of data packets by the addition of redundant bits in coding stage. The addition of bits helps to achieve the reliability of transmission, but it may result in consumption of extra power at nodes.

It is evident from the discussions above that the security and reliability form the major challenges in the WSN of current times. Both security and reliability are to be tackled from the internal and external aspects.

Externally the physical devices are always vulnerable to vandalism and damage by attackers. The devices require better physical protection and privacy with techniques like casing, sealing, deployment at unreachable locations and hidden placement. The external attacks are subjected to all layers of the WSN. The layer must be equipped with TRM system, which can facilitate easier and reliable communication among the nodes of the WSN. The authentication must be bidirectional and must contain the maximum bit length allowed. The access control levels must be made mandatory and be defined clearly to avoid the access by unauthorized personnel. The global and the public keys must be refreshed and revalidated more frequently. The algorithm to generate the keys is required to be the most well-protected part of the system. The data must be encrypted in such a way that the topology and the environment are hidden from the attacker's view.

Internally, most of the time, security becomes a threat due to the betrayal of the internal employees of the organization. Thus, it is very essential to maintain an activity log of the current employees and performance of background check of the new recruits. Creation of awareness of the security issues, countermeasures and best practices among individuals of the organization requires more attention and initiative. The managerial, operational and the technical domains of the WSN must always form an interlink to fight the attacks. The reliability of the hardware and software must be validated with extensive testing. The testing must also be performed during runtime to verify the integrity of the hardware and the software.

3.4 Conclusion

This chapter provides a detailed survey of the security issues with special emphasis on the trust and reputation mechanisms. The security issues, attacks, features, and consequences of the attacks and some of the countermeasures have been discussed. A systematic literature review has been conducted on the concept of trust and reputation in WSNs and the report is arranged in the yearly timeline. The future scope of the concept of trust and reputation have been identified and discussed. Some of the shortcomings, challenges in security, and reliability are also examined.

Although various methods for security have been developed and implemented throughout the years, based on the literature and its understanding, it can be anticipated that the trust and reputation mechanism forms the best security technique for WSNs. This is due to the simplicity and the effectiveness of the technique. It also eliminates some of the time-consuming and complex processes, which are almost infeasible due to the low features available in the nodes.

References

Abouei, J., Brown, J. D., Plataniotis, K. N., & Pasupathy, S. (2011). Energy efficiency and reliability in wireless biomedical implant systems. *IEEE Transactions on Information Technology in Biomedicine, 15,* 456–466.

Almasri, M., Elleithy, K., Bushang, A., & Alshinina, R. (2013). Terp: A trusted and energy efficient routing protocol for wireless sensor networks (WSNs). In *Proceedings of the 2013 IEEE/ACM 17th International Symposium on Distributed Simulation and Real Time Applications, 2013* (pp. 207–214). IEEE Computer Society.

Amini, F., Mišic, V. B., & Mišic, J. (2007). Intrusion detection in wireless sensor networks. *Security in Distributed, Grid, Mobile, and Pervasive Computing,* pp. 111

Anand, M., Cronin, E., Sherr, M., Blaze, M., Ives, Z., & Lee, I. (2006). Security challenges in next generation cyber physical systems. In *Beyond SCADA: Networked Embedded Control for Cyber Physical Systems*

Anderson, R., Chan, H., & Perrig, A. (2004). Key infection: Smart trust for smart dust. In *ICNP 2004. Proceedings of the 12th IEEE International Conference on Network Protocols, 2004* (pp. 206–215). New York: IEEE.

Araujo, A., Blesa, J., Romero, E., & Villanueva, D. (2012). Security in cognitive wireless sensor networks. Challenges and open problems. *EURASIP Journal of Wireless Communications and Networking, 2012,* 48.

Ashraf, N., Bohnet, I., & Piankov, N. (2006). Decomposing trust and trustworthiness. *Experimental Economics, 9,* 193–208.

Baig, W. A., Khan, F. I., Kim, K.-H., & Yoo, S.-W. (2012). Privacy assuring protocol using simple cryptographic operations for smart metering. *International Journal of Multimedia and Ubiquitous Engineering, 7,* 315–322.

Bao, F., Chen, I.-R., Chang, M., & Cho, J.-H. (2011a). Hierarchical trust management for wireless sensor networks and its application to trust-based routing. In *Proceedings of the 2011 ACM Symposium on Applied Computing* (pp. 1732–1738). ACM, New York.

Bao, F., Chen, R., Chang, M., & Cho, J.-H. (2011b). Trust-based intrusion detection in wireless sensor networks. In *2011 IEEE International Conference on Communications (ICC)* (pp. 1–6). New York: IEEE.

Bao, F., Chen, R., Chang, M., & Cho, J.-H. (2012). Hierarchical trust management for wireless sensor networks and its applications to trust-based routing and intrusion detection. *IEEE Transactions on Network and Service Management, 9,* 169–183.

Becher, A., Benenson, Z., & Dornseif, M. (2006). Tampering with Motes: Real-world physical attacks on wireless sensor networks. In J. A. Clark, R. F. PAIGE, F. A. C. Polack, & P. J. Brooke (Eds.), *Security in Pervasive Computing: Third International Conference, SPC 2006, York, UK, April 18–21, 2006. Proceedings.* Berlin, Heidelberg: Springer Berlin Heidelberg.

Benenson, Z., Cholewinski, P. M., & Freiling, F. C. (2008). Vulnerabilities and attacks in wireless sensor networks. *Wireless Sensors Networks Security,* pp. 22–43.

Blaze, M., Feigenbaum, J., & Lacy, J. (1996). Decentralized trust management. In *IEEE Symposium on Security and Privacy* (pp. 164–173). New York: IEEE.

Blilat, A., Bouayad, A., El Houda Chaoui, N., & Ghazi, M. (2012). Wireless sensor network: Security challenges. In *2012 National Days of Network Security and Systems (JNS2), 2012* (pp. 68–72). New York: IEEE.

Boukerch, A., Xu, L., & El-Khatib, K. (2007). Trust-based security for wireless ad hoc and sensor networks. *Computer Communications, 30,* 2413–2427.

Boukerche, A., Oliveira, H., Nakamura, E. F., & Loureiro, A. A. (2008). Secure localization algorithms for wireless sensor networks. *IEEE Communications Magazine, 46,* 96–101.

Byers, J., & Nasser, G. (2000). Utility-based decision-making in wireless sensor networks. In *2000 First Annual Workshop on Mobile and Ad Hoc Networking and Computing, 2000. MobiHOC* (pp. 143–144). New York: IEEE.

Cardenas, A., Amin, S., & Sastry, S. (2008). Secure control towards survivable cyber-physical systems. In *28th International Conference on Distributed Computing Systems Workshops, 2008. ICDCS '08*. Beijing: IEEE.

Che, S., Feng, R., Liang, X., & Wang, X. (2015). A lightweight trust management based on Bayesian and Entropy for wireless sensor networks. *Security and Communication Networks, 8,* 168–175.

Chen, H., Gu, G., Wu, H., & Gao, C. (2007a). Reputation and trust mathematical approach for wireless sensor networks. *International Journal of Multimedia and Ubiquitous Engineering,* pp. 23–32.

Chen, H., Wu, H., Zhou, X., & Gao, C. (2007b). Reputation-based trust in wireless sensor networks. In *International Conference on Multimedia and Ubiquitous Engineering, 2007. MUE'07* (pp. 603–607). New York: IEEE.

Choi, J.-H., Choi, C., Ko, B.-K., & Kim, P.-K. (2012). Detection of cross site scripting attack in wireless networks using n-Gram and SVM. *Mobile Information Systems, 8,* 275–286.

Cinque, M., Cotroneo, D., De Caro, G., & Pelella, M. (2006). Reliability requirements of wireless sensor networks for dynamic structural monitoring. In *International Workshop on Applied Software Reliability (WASR 2006)* (pp. 8–13).

Cowan, C., Wagle, P., Pu, C., Beattie, S., & Walpole, J. (2000). Buffer overflows: Attacks and defenses for the vulnerability of the decade. In *Proceedings of DARPA Information Survivability Conference and Exposition, 2000. DISCEX'00* (pp. 119–129). New York: IEEE.

Crosby, G. V., Hester, L., & Pissinou, N. (2011). Location-aware, trust-based detection and isolation of compromised nodes in wireless sensor networks. *IJ Network Security, 12,* 107–117.

Crosby, G. V., & Pissinou, N. (2007). Cluster-based reputation and trust for wireless sensor networks. In *Consumer Communications and Networking Conference*

Dhulipala, V. S., Karthik, N., & Chandrasekaran, R. (2013). A novel heuristic approach based trust worthy architecture for wireless sensor networks. *Wireless Personal Communications, 70,* 189–205.

Du, X., Xiao, Y., Chen, H. H., & Wu, Q. (2006). Secure cell relay routing protocol for sensor networks. *Wireless Communications and Mobile Computing, 6,* 375–391.

Egan, D. (2005). The Emergence of ZigBee in building automation and industrial controls. *Computing and Control Engineering, 16,* 14–19.

Estrin, D., Govindan, R., Heidemann, J., & Kumar, S. (1999). Next century challenges: Scalable coordination in sensor networks. In *Proceedings of the 5th Annual ACM/IEEE International Conference on Mobile Computing and Networking* (pp. 263–270). New York: ACM.

Fatema, N., & Brad, R. (2014). Attacks and counterattacks on wireless sensor networks. *arXiv preprint* arXiv:1401.4443.

Fernandez-Gago, M. C., Roman, R., & Lopez, J. (2007). A survey on the applicability of trust management systems for wireless sensor networks. In *Third International Workshop on Security, Privacy and Trust in Pervasive and Ubiquitous Computing, 2007. SECPerU 2007* (pp. 25–30). New York: IEEE.

Fraley, C. (1998). Algorithms for model-based Gaussian hierarchical clustering. *SIAM Journal on Scientific Computing, 20,* 270–281.

Franzèn, J. (2008). *Bayesian cluster analysis.* Ph.D. Doctrate, Stockholm.

Ganeriwal, S., Balzano, L. K., & Srivastava, M. B. (2008). Reputation-based framework for high integrity sensor networks. *ACM Transactions on Sensor Networks (TOSN), 4,* 15.

Govindan, K., & Mohapatra, P. (2012). Trust computations and trust dynamics in mobile adhoc networks: A survey. *IEEE Communications Surveys & Tutorials, 14,* 279–298.

Han, G., Jiang, J., Shu, L., Niu, J., & Chao, H.-C. (2014). Management and applications of trust in Wireless Sensor Networks: A survey. *Journal of Computer and System Sciences, 80,* 602–617.

Hardin, R. (2002). *Trust and trustworthiness.* Russell Sage Foundation.

Hsu, M.-T., Lin, F. Y.-S., Chang, Y.-S., & Juang, T.-Y. (2007). The reliability of detection in wireless sensor networks: Modeling and analyzing. In *Embedded and ubiquitous computing.* Berlin: Springer.

Huang, S.-I., Shieh, S., & Tygar, J. (2010). Secure encrypted-data aggregation for wireless sensor networks. *Wireless Networks, 16,* 915–927.

Jadidoleslamy, H. (2014). A comprehensive comparison of attacks in wireless sensor networks. *International Journal of Computer Communications and Networks (IJCCN), 4.*

Jiang, T., & Baras, J. S. (2004). Ant-based adaptive trust evidence distribution in MANET. In *Proceedings. 24th International Conference on Distributed Computing Systems Workshops, 2004* (pp. 588–593). New York: IEEE.

Jsang, A., & Ismail, R. (2002). The beta reputation system. In *Proceedings of the 15th Bled Electronic Commerce Conference* (pp. 41–55).

Kamvar, S. D., Schlosser, M. T., & Garcia-Molina, H. (2003). The Eigentrust algorithm for reputation management in p2p networks. In *Proceedings of the 12th International Conference on World Wide Web, 2003* (pp. 640–651). New York: ACM.

Karthik, N., & Karthik, J. (2014). Trust worthy framework for wireless sensor networks. *International Journal of Computer Science & Engineering Technology (IJCSET), 5,* 478–480.

Kaur, D., & Singh, P. (2014). Various OSI layer attacks and countermeasure to enhance the performance of WSNs during wormhole attack. *International Journal on Network Security, 5* (1), 6.

Khalid, O., Khan, S. U., Madani, S. A., Hayat, K., Khan, M. I., Min-Allah, N., et al. (2013). Comparative study of trust and reputation systems for wireless sensor networks. *Security and Communication Networks, 6,* 669–688.

Kim, T. K., & Seo, H. S. (2008). A trust model using fuzzy logic in wireless sensor network. *World Academy of Science, Engineering and Technology, 42,* 63–66.

Kinney, P. (2003). Zigbee technology: Wireless control that simply works. In *Communications Design Conference* (pp. 1–7).

Kumar, V., Jain, A., & Barwal, P. (2014). Wireless sensor networks: Security issues, challenges and solutions. *International Journal of Information & Computation Technology,* pp. 0974–2239.

Lee, K. (2000). IEEE 1451: A standard in support of smart transducer networking. In *Instrumentation and Measurement Technology Conference, 2000. IMTC 2000. Proceedings of the 17th IEEE, 2000* (pp. 525–528). New York: IEEE.

Lewis, F. L. (2004). *Wireless sensor networks, smart environments: Technologies, protocols, and applications* (pp. 11–46).

Li, X., Zhou, F., & Du, J. (2013). LDTS: A lightweight and dependable trust system for clustered wireless sensor networks. *IEEE Transactions on Information Forensics and Security,* pp. 924–935.

Li, Z., & Gong, G. (2008). Survey on security in wireless sensor. *Special English Edition of Journal of KIISC, 18,* 233–248.

Liu, A. X., Kovacs, J. M., Huang, C.-T., & Gouda, M. G. A secure cookie protocol. In *Proceedings of 14th International Conference on Computer Communications and Networks, 2005,* San Diego, California, USA (pp. 333–338). New York: IEEE.

Liu, Z., Joy, A. W., & Thompson, R. A. (2004). A dynamic trust model for mobile ad hoc networks. In *Proceedings of the 10th IEEE International Workshop on Future Trends of Distributed Computing Systems, 2004. FTDCS 2004* (pp. 80–85). New York: IEEE.

Lopez, J., Roman, R., Agudo, I., & Fernandez-Gago, C. (2010). Trust management systems for wireless sensor networks: Best practices. *Computer Communications, 33,* 1086–1093.

Lopez, J., Roman, R., & Alcaraz, C. (2009). Analysis of security threats, requirements, technologies and standards in wireless sensor networks. In *Foundations of security analysis and design V.* Berlin: Springer.

López, J., & Zhou, J. (2008). *Wireless sensor network security.* Amsterdam: IOS Press.

Lynch, C. (1999). Canonicalization: A fundamental tool to facilitate preservation and management of digital information. *D-Lib Magazine, 5,* 17–25.

Mahmood, M. A., Seah, W. K., & Welch, I. (2015). Reliability in wireless sensor networks: A survey and challenges ahead. *Computer Networks, 79,* 166–187.

Mármol, F. G., & Pérez, G. M. (2009). TRMSim-WSN, trust and reputation models simulator for wireless sensor networks. In IEEE International Conference on Communications, 2009 (pp. 1–5). ICC'09. New York: IEEE.

Mármol, F. G., & Pérez, G. M. (2010). Towards pre-standardization of trust and reputation models for distributed and heterogeneous systems. *Computer Standards & Interfaces, 32,* 185–196.

Miao, C.-Y., Dai, G.-Y., & Chen, Q.-Z. (2015). Cooperative localization and location verification in WSN. In *Human centered computing.* Berlin: Springer.

Mohammadi, S., & Jadidoleslamy, H. (2011a). A comparison of link layer attacks on wireless sensor networks. *arXiv preprint* arXiv:1103.5589.

Mohammadi, S., & Jadidoleslamy, H. (2011b). A comparison of physical attacks on wireless sensor networks. *International Journal of Peer to Peer Networks, 2,* 24–42.

Momani, M., Aboura, K., & Challa, S. (2007a). RBATMWSN: Recursive Bayesian approach to trust management in wireless sensor networks.

Momani, M., & Challa, S. (2008). GTRSSN: Gaussian trust and reputation system for sensor networks. In *Advances in computer and information sciences and engineering.* Berlin: Springer.

Momani, M., & Challa, S. (2010). Survey of trust models in different network domains. *International Journal of Ad Hoc, Sensor & Ubiquitous Computing, 1,* 1–19.

Momani, M., Challa, S., & Aboura, K. (2007b). Modelling trust in wireless sensor networks from the sensor reliability prospective. In *Innovative algorithms and techniques in automation, industrial electronics and telecommunications.* Berlin: Springer

Momani, M., Challa, S., & Alhmouz, R. (2008). Can we trust trusted nodes in wireless sensor networks? In *International Conference on Computer and Communication Engineering, 2008. ICCCE 2008* (pp. 1227–1232). New York: IEEE.

Naseer, A. (2012). *Reputation system based trust-enabled routing for wireless sensor networks.* INTECH Open Access Publisher.

Ngai, E.-H., & Lyu, M. R. (2004). Trust-and clustering-based authentication services in mobile ad hoc networks. In *Proceedings of 24th International Conference on Distributed Computing Systems Workshops, 2004* (pp. 582–587). New York: IEEE.

Patil, V. S., Bamnote, G. R., & Nair, S. S. (2011). Cross site scripting: An overview. In *IJCA Proceedings on International Symposium on Devices MEMS, Intelligent Systems and Communication*, (pp. 19–22), Sikkim, India.

Pereira, P. R., Grilo, A., Rocha, F., Nunes, M. S., Casaca, A., Chaudet, C., et al. (2007). End-to-end reliability in wireless sensor networks: Survey and research challenges. In *EuroFGI Workshop on IP QoS and Traffic Control* (pp. 67–74).

Pirzada, A. A., & Mcdonald, C. (2004). Establishing trust in pure ad-hoc networks. In *Proceedings of the 27th Australasian Conference on Computer science-Volume 26, 2004* (pp. 47–54). Australian Computer Society, Inc.

Probst, M. J., & Kasera, S. K. (2007). Statistical trust establishment in wireless sensor networks. In *2007 International Conference on Parallel and Distributed Systems* (pp. 1–8). New York: IEEE.

Reddy, Y., & Selmic, R. (2011). Agent-based trust calculation in wireless sensor networks. In *SENSORCOMM 2011, The Fifth International Conference on Sensor Technologies and Applications* (pp. 334–339).

Rehana, J. (2009). Security of wireless sensor network. *Helsinki University of Technology, Technical report*.

Reshmi, V., & Sajitha, M. (2014). A survey on trust management in wireless sensor networks. *International Journal of Computer Science & Engineering Technology, 5*, 104–109.

Sabeel, U., Maqbool, S., & Chandra, N. (2013). Categorized security threats in the wireless sensor networks: Countermeasures and security management schemes. *International Journal of Computer Applications, 64*, 19–28.

Sarma, H. K. D., & Kar, A. (2006) Security threats in wireless sensor networks. In *Proceedings 2006 40th Annual IEEE International, Carnahan Conferences Security Technology* (pp. 243–251). New York: IEEE.

Sarma, H. K. D., & Kar, A. (2008). Security threats in wireless sensor networks. *IEEE Aerospace and Electronic Systems Magazine, 23*, 39–45.

Sastry, A. S., Sulthana, S., & Vagdevi, S. (2013). Security threats in wireless sensor networks in each layer. *International Journal of Advanced Networking and Applications, 4*, 1657.

Shaikh, R. A., Jameel, H., D'Auriol, B. J., Lee, H., Lee, S., & Song, Y.-J. (2009). Group-based trust management scheme for clustered wireless sensor networks. *IEEE Transactions on Parallel and Distributed Systems, 20*, 1698–1712.

Shannon, C. (1948). A mathematical theory of communication. *The Bell System Technical Journal, 27*, 379–423 & 623–656.

Silberman, P., & Johnson, R. (2004). A comparison of buffer overflow prevention implementations and weaknesses. In *IDEFENSE, August*.

Singh, H., Agrawal, M., Gour, N., & Hemrajani, N. (2014). A study on security threats and their countermeasures in sensor network routing. *Prevention, 3*.

Singh, M., Sardar, A. R., Sahoo, R. R., Majumder, K., Ray, S., & Sarkar, S. K. (2015). Lightweight trust model for clustered WSN. In *Proceedings of the 3rd International Conference on Frontiers of Intelligent Computing: Theory and Applications (FICTA) 2014* (pp. 765–773). Berlin: Springer.

Sorniotti, A., Gomez, L., Wrona, K., & Odorico, L. (2007). Secure and trusted in-network data processing in wireless sensor networks: A survey. *Journal of Information Assurance and Security, 2*, 189–199.

Srinivasan, A., Teitelbaum, J., & Wu, J. (2006). DRBTS: Distributed reputation-based beacon trust system. In *2nd IEEE International Symposium on Dependable, Autonomic and Secure Computing, 2006* (pp. 277–283). New York: IEEE.

Srinivasan, A., Teitelbaum, J., Wu, J., Cardei, M., & Liang, H. (2009). Reputation-and-trust-based systems for ad hoc networks. In *Algorithms and protocols for wireless and mobile ad hoc networks*, pp. 375.

Srivastava, S., & Johri, K. (2012). A survey on reputation and trust management in wireless sensor network. *International Journal of Scientific Research Engineering & Technology, 1*, 139–149.

Sun, F., Zhao, Z., Fang, Z., Du, L., Xu, Z., & Chen, D. (2014). A review of attacks and security protocols for wireless sensor networks. *Journal of Networks, 9*, 1103–1113.

Undercoffer, J., Avancha, S., Joshi, A., & Pinkston, J. (2002). Security for sensor networks. In *CADIP Research Symposium*. Citeseer (pp. 25–26).

Virmani, D., Soni, A., Chandel, S., & Hemrajani, M. (2014). Routing attacks in wireless sensor networks: A survey. *arXiv preprint* arXiv:1407.3987.

Wang, C., Sohraby, K., Li, B., Daneshmand, M., & Hu, Y. (2006a). A survey of transport protocols for wireless sensor networks. *IEEE Network, 20*, 34–40.

Wang, C., Sohraby, K., Li, B., & Tang, W. (2005). Issues of transport control protocols for wireless sensor networks. In *Proceedings of International Conference on Communications, Circuits and Systems (ICCCAS)* (pp. 422–426).

Wang, N., Gao, L., & Wu, C. (2014). A light-weighted data trust model in WSN. *International Journal of Grid & Distributed Computing, 7*.

Wang, Y., Attebury, G., & Ramamurthy, B. (2006b). A survey of security issues in wireless sensor networks. *IEEE Communications Surveys & Tutorials, 8,* 2–23.

Willig, A., & Karl, H. (2005). Data transport reliability in wireless sensor networks. A survey of issues and solutions. *Praxis der Informationsverarbeitung und Kommunikation, 28,* 86–92.

Wright, S. (2010). Trust and trustworthiness. *Philosophia, 38,* 615–627.

Xing, K., Srinivasan, S. S. R., Jose, M., Li, J., & Cheng, X. (2010). Attacks and countermeasures in sensor networks: A survey. In *Network security.* Berlin: Springer

Yang, S.-H. (2014). *Wireless sensor networks, principles, design and applications.* London: Springer.

Yang, S.-H., & Cao, Y. (2008). Networked control systems and wireless sensor networks: Theories and applications.

Yick, J., Mukherjee, B., & Ghosal, D. (2008). Wireless sensor network survey. *Computer Networks, 52,* 2292–2330.

Yin, J., & Madria, S. K. (2006). SecRout: A secure routing protocol for sensor networks. In *20th International Conference on Advanced Information Networking and Applications, 2006. AINA 2006* (pp. 6). New York: IEEE.

Yu, H., Shen, Z., Miao, C., Leung, C., & Niyato, D. (2010). A survey of trust and reputation management systems in wireless communications. *Proceedings of the IEEE, 98,* 1755–1772.

Yu, Y., Li, K., Zhou, W., & Li, P. (2012). Trust mechanisms in wireless sensor networks attack analysis and countermeasures. *Journal of Network and Computer Application, 35,* 867–880.

Zhan, G., Shi, W., & Deng, J. (2009). SensorTrust: A resilient trust model for WSNs. In *Proceedings of the 7th ACM Conference on Embedded Networked Sensor Systems, 2009* (pp. 411–412). New York: ACM.

Zhang, W., Das, S. K., & Liu, Y. (2006a). A trust based framework for secure data aggregation in wireless sensor networks. In *2006 3rd Annual IEEE Communications Society on Sensor and Ad Hoc Communications and Networks, 2006. SECON'06* (pp. 60–69). New York: IEEE.

Zhang, Y., Liu, W., Fang, Y., & Wu, D. (2006b). Secure localization and authentication in ultra-wideband sensor networks. *IEEE Journal on Selected Areas in Communications, 24,* 829–835.

Zia, T., & Zomaya, A. (2006). Security issues in wireless sensor networks. In *International Conference on Systems and Networks Communications, 2006. ICSNC'06* (pp. 40–40). New York: IEEE.

Chapter 4
WSN Security Mechanisms for CPS

Due to their low complexity and robustness in nature, wireless sensor networks are a key component in cyber-physical system. The integration of wireless sensor network in cyber-physical system provides immense benefits in distributed controlled environment. However, the layered structure of cyber-physical system and wireless sensor network make it susceptible to internal and external threats. These threats may lead toward financial or structural losses in networks. The chapter is structured as such to provide classification of layer-to-layer, external and internal attacks to wireless sensor network and cyber-physical system. In addition to that, the chapter identifies the known security detection and possible approaches against the threats for wireless sensor networks and cyber-physical system. Finally, a comparison of approaches to defend wireless sensor network and cyber-physical system against such attacks is presented.

4.1 Introduction

Advancement in the fields of embedded systems has enabled the utilization of low-cost wireless sensor networks (WSN) in cyber-physical system (CPS). CPS are large, heterogeneous distributed control systems with actuators and sensors that enable interactions between human to machine, machine to human, and human to object communication in cyber or physical world (Ali et al. 2015). To achieve the wireless communication between such platforms, CPS adopts wireless sensors networks as it being a heterogeneous module, also has the capability to operate autonomously in any environment. The function of WSN is to collect data from neighboring nodes at aggregation point and forward it to the base station to process the gathered data (Xing et al. 2010a, b). The gathered data can be from any domestic application, an engineering application, or from any battlefield that monitors the activities of the enemy. However, with its limited resources, small

© Springer International Publishing AG 2018
S. Ali et al., *Cyber Security for Cyber Physical Systems*, Studies in Computational Intelligence 768, https://doi.org/10.1007/978-3-319-75880-0_4

performance capabilities and applications in unattended environments, WSNs face a wide range of security issues (Han et al. 2014).

Cyber-physical systems are defined as a system which is an integration of computational, networking, control and physical processes (Stelte and Rodosek 2013; Kim and Kumar 2013). The physical and the cyber parts work closely to sense the changes in the real world and take necessary steps to obtain the desired results (Lu et al. 2014). The physical processes are monitored and controlled by the cyber systems, which are small devices similar to the ones in WSNs (Shafi 2012). The physical devices are mainly the sensors and the actuators. CPS has inherited many of the security issues from its ancestors and has created a new set of issues derived from its advancement. However, there are threats associated with each level of such models which this chapter will discuss further in the following sections.

4.2 Related Work

As mentioned in the previous chapters, CPS is a technology majorly developed from the WSN (Wood et al. 2009; Wan et al. 2010, 2013b). Various efforts have been taken into account to identify the similarities and differences between the two fields and activities required to transform from WSN to CPS. Wu et al. (2011) analyzed challenges that need to be overcome for moving from WSN toward the CPS. The study has a uniqueness to distinguish between WSN, MANET, and CPS in terms of network formation, power management, and communication. In addition to that, the research also provides a limited insight into the security issues involved. Xia et al. (2011) identified the lack of attention toward the security of cyber-physical system. Mao et al. (2011) have categorized the issues of security in the field of WSN, CPS and Internet of Things (IoT). Lin et al. (2012) discussed the importance of WSN component for CPS and the challenges involved in the integration process. The study is based on the available literature and provides the overview of the architecture and features of WSN to be used in CPS. The study has no uniqueness as it does not provide any viable solution for security. Wan et al. (2013a) studied the transformation of M2M and WSN to CPS. The study emphasized toward the need of security but failed to identify their features, categories, and countermeasures. Ali et al. (2015) studied the utilization of WSN in CPS with various dimensions of security requirement in CPS system and their parameters. The study has no uniqueness as it discusses the requirements of cyber-physical system but fails to provide any solution for that as well.

As observed in the existing literature the issue of security has initially received less significance from the earlier development of the technology. With the advancement of time, the flaws became more evident due to threats and vulnerabilities. This led to greater emphasis on the security concerns in this field.

4.3 Security Concerns

Taking from the perception demonstrated in (Wan et al. 2013a), CPS and WSN, both come under the common umbrella of IoT. Thus, most of the issues and concerns including that of security are similar. The major difference is that the CPS is a system at larger scale which combines the technologies of WSN and M2M. A single CPS could contain multiple clusters and hierarchies of WSNs. Alcaraz and Lopez (2015), Mahmood et al. (2015) and Wang et al. (2006) accentuated that wireless sensor network is susceptible to various kinds of attacks. The major point of concern is due to the limited resources, which makes integration of high-level security techniques impractical in the low level of memory and processors available. In addition to that, Li and Gong (2008), Yu et al. (2012), Lopez et al. (2009) considered that the security issues are related to confidentiality, integrity, authenticity authorization, survivability, secrecy, availability, scalability, efficiency, and confidentiality. In cyber-physical system, the security issues are not only inherited from WSN but also has security issues extends to elements like actuators, decision-making computations and the cross-domain communication, and the heterogeneous information flow in both directions. Cardenas et al. (2008), Lu et al. (2014), Shafi (2012), Pal et al. (2009), Ali and Anwar (2012), Saqib et al. (2015) and Muhammad Farhan (2014) identified that major issues related to security in cyber-physical system are confidentiality availability, authenticity, validation, integrity, reliability, robustness, and trustworthiness. Saqib et al. (2015), Ali and Anwar (2012), Govindarasu et al. (2012), Aloul et al. (2012) identified that the above mentioned issues leads to various security threats that are denial of service, man-in-the-middle, time synchronization attack, routing attacks, malware, network-based intrusion, eavesdropping, compromised key attack, resonance attack, integrity attack and jamming attack.

The comparison of layer-to-layer attacks between wireless sensor network and cyber-physical system will be discussed in Sect. 4.4.

4.4 Layer-to-Layer Attacks Between CPS and WSN

CPS and WSN are structured in layered architecture; the layered architecture of this network is more exposed to vulnerabilities and could lead to incredible harm and losses due to a number of different attacks as discussed in the previous sections. In this section, the layers of CPS and WSN will be presented and their various detection techniques for recognition of attack and providing active or passive defensive mechanism for the prevention of network will be discussed as well.

4.4.1 External Attacks on Physical Layer

The physical layer of CPS and WSN is susceptible to different attacks, such as traffic analysis attack, eavesdropping, frequency jamming, device tempering, Sybil, and path-based DoS attacks. Table 4.1 illustrates the effects of attack on physical layer of WSN and CPS with their detection and defensive mechanism. Deng et al. (2004) classified traffic analysis attack in WSN in two different categories, rate monitoring attack and time correlation attack. The rate monitoring attack deals with the number of packets sent by the nodes toward the adversary. The time correlation attack deals with correlation between the sending times of neighboring nodes. In addition to that, a threat model is also introduced for the purpose of simulation of rate monitoring attack by different anti-traffic analysis techniques whereas the efficiency of time correlation attack is mitigated by introducing the random fake paths. The techniques introduced by Deng et al. (2004), based on multi-parent routing scheme (MPR), random walk (RW), differential fractal propagation (DFP), fractal propagation with different forking probabilities (DEFP), and enforced fractal propagation (EFP), which has uniqueness that it hides the location of base station but all these techniques are not practically implemented on any sensor network. On the contrary, Alcaraz and Lopez (2015) in SCADA networks identified that Zigbee pro is a more viable approach toward the traffic analysis attack, as it selects random path for communication.

Jadidoleslamy (2014) and Virmani et al. (2014) suggested strong encryption technique, systematic access control and advanced encryption as a proposed solution for the eavesdropping in wireless sensor network. Shin et al. (2010) conducted an experiment by using one hop clustering for SCADA application by using the existing WSN techniques. The protocol developed to be helpful to prevent bogus routing information data and sinkhole attack. The proposed solution by Shin et al. (2010) is the experimental study of the previously identified intrusion detection techniques but not viable for heterogonous network. Another threat is interference in intensive radio signals which has been studied by a number of authors (Mohammadi and Jadidoleslamy 2011b; Virmani et al. 2014; Sabeel et al. 2013). The authors proposed various solutions to counter the effect of jamming, that are secure encryption, CRC check, lower duty cycle, higher broadcast power, hybrid FHSS/DSSS, ultra-wideband, change of antenna polarization, and use of directional transmission. In CPS, Mo et al. (2012) suggests the use of different spread spectrum techniques for the prevention of Jamming attacks. However, that study is not comprehensive, as it just provides the overview of the available techniques, whereas the suggested model can only evaluate the replay attack through simulation. Mohammadi and Jadidoleslamy (2011b) studied the network degradation and damages in the functionality of the sensor nodes by P-DoS attack. In addition to that, they also studied the effects of P-DoS attack that can be over-come by using redundancy of network packet data, acknowledge validation of receiving data, and gray-listing of compromised nodes. Shin et al. (2010) studied

Table 4.1 Comparison of attacks on WSN and CPS

Attack	Effects	WSN		CPS	
		Detection methods	Defensive methods	Detection methods	Defensive methods
External attacks in physical layer					
Traffic analysis	Deteriorated network performance, high packet collision, traffic alteration (Deng et al. 2004)	Statistical analysis of typical rate monitoring and time correlation attack (Deng et al. 2004)	Multi-parent routing scheme, random walk, fractal propagation (Deng et al. 2004)	Statistical analysis (Alcaraz and Lopez 2015)	Model base data traffic analysis by Modbus TCP in Industrial wireless sensor network (Alcaraz and Lopez 2015)
Eavesdropping pay close vigilance to the communication channel	Reduction in data privacy, Extraction of vital data. Exposure to adversaries Secondary attacks (Virmani et al. 2014; Mohammadi and Jadidoleslamy 2011b)	Statistical analysis, Misbehavior detection techniques (Virmani et al. 2014)	Systematic access control, dispersed processing, access constraint, advanced encryption, peripatetic security solution, strong encryption technique (Virmani et al. 2014; Mohammadi and Jadidoleslamy 2011b)	Behavior, behavior-specification, knowledge (Mitchell and Chen 2014)	Shin technique (Shin et al. 2010)
Jamming: cause interference by introducing intensive radio signals of identical frequency	Excessive energy, Disrupt communication, Occupy entire bandwidth, Corrupt data packets, Deceive network's defensive mechanisms, Resource exhaustion (Mohammadi and Jadidoleslamy 2011b; Virmani et al. 2014; Sabeel et al. 2013)	Statistical information, threshold for Channel utility degradation, Background noise (Mohammadi and Jadidoleslamy 2011b; Virmani et al. 2014; Sabeel et al. 2013)	Secure Encryption, CRC check, Lower duty cycle, Higher broadcast power, Hybrid FHSS/DSSS, Ultra-wideband, Change of antenna Polarization, Use of directional Transmission, Message prioritization, Blacklist (Mohammadi and Jadidoleslamy 2011b; Virmani et al. 2014; Sabeel et al. 2013)	Statistical Information	spread spectrum techniques, FHSS/DSSS, encryption (Mo et al. 2012)
Device tampering: direct physical capture of sensor. Attack at the base station	Captivate & destroy captured node to clone it to annex WSN using software liabilities (Sabeel et al. 2013; Sun et al. 2014b; Mohammadi and Jadidoleslamy 2011b)	Intermodal isolation, monitoring, key management, misbehavior (Sabeel et al. 2013; Sun et al. 2014b; Mohammadi and Jadidoleslamy 2011b)	Hardware and software alertness, Disguising sensors, Restriction access, Data reliability and privacy, Node detection (Sabeel et al. 2013; Sun et al. 2014b; Mohammadi and Jadidoleslamy 2011b)	Intermodal isolation, monitoring, Key management, misbehavior (Networks 2016)	Host Identity Protocol (HIP): A New Trust Model (Networks 2016)

(continued)

Table 4.1 (continued)

Attack	Effects	WSN		CPS	
		Detection methods	Defensive methods	Detection methods	Defensive methods
Sybil attack: system sabotaged by counterfeiting characters	Causes network inaccessibility (Mohammadi and Jadidoleslamy 2011b; Sabeel et al. 2013)	Low overhead and delay of signals (Mohammadi and Jadidoleslamy 2011b; Sabeel et al. 2013)	Physical shield of nodes (Mohammadi and Jadidoleslamy 2011b; Sabeel et al. 2013)	Low overhead and delay of signals (Newsome et al. 2004)	Radio Resource testing, Random key redistribution, Registration, position verification, code attestation (Newsome et al. 2004)
Path-based DoS: typical mixed jamming like attacks	Exhaustion of nodes' battery, network disturbance, deceitful rejection of nodes (Mohammadi and Jadidoleslamy 2011b)	Misbehavior (Mohammadi and Jadidoleslamy 2011b)	Redundancy, anti-attack, acknowledge validation, gray-listing (Mohammadi and Jadidoleslamy 2011b)	Behavior, behavior-specification (Shin et al. 2010), knowledge of packet arrival rate, message type (Shin et al. 2010)	IDS: Shin, Cheung, Gao, Yang (Shin et al. 2010)
External attacks in data link layer					
Collision	Interferences, corruption in Data or control packets. Packets discard. Energy Over-usage. Effect on cost (Sun et al. 2014b; Mohammadi and Jadidoleslamy 2011a)	Misbehavior	Error correction codes, Time diversity (Sun et al. 2014b; Mohammadi and Jadidoleslamy 2011a)	Statistical analysis (Gill 2002)	Error correcting codes (Gill 2002)
Resource exhaustion	• Resources exhaustion • Compromise availability (Fatema and Brad 2013; Mohammadi and Jadidoleslamy 2011a)	Misbehavior (Fatema and brad 2013; Mohammadi and Jadidoleslamy 2011a)	Restrictive MAC rate. Back-offs at a random rate. TDM. Regulating link response rate. ID protection (Fatema and brad 2013; Mohammadi and Jadidoleslamy 2011a)	Misbehavior (Gill 2002)	Real-time monitoring mechanism, Correlating event logs, related to IED relay settings (Aniket 2009) Rate Limitation (Gill 2002)
Traffic manipulation	Aggressive channel usage, WSN ineffectualness, traffic falsification, artificially contention, deteriorated signal quality (Sabeel et al. 2013)	Misbehavior (Sabeel et al. 2013; Virmani et al. 2014)	Traffic analysis, Collision defenses, Link Layer encryption. Rate of MAC requests (Sabeel et al. 2013; Virmani et al. 2014)	Behavior, behavior-specification, knowledge (Aniket 2009)	Use of Network management protocol such as SNMP (Aniket 2009)

(continued)

Table 4.1 (continued)

Attack	Effects	WSN		CPS	
		Detection methods	Defensive methods	Detection methods	Defensive methods
Eavesdropping	Extract critical data, reduced privacy protection (Jadidoleslamy 2014)	Statistical analysis	Control of access. Distributed processing. High encryption (Jadidoleslamy 2014)	Statistical analysis	Shin Technique (Shin et al. 2010)
Impersonation: killing of the cluster leader and diverting nodes to a wrong position	Disrupted routing tables, kill sensors. Congested network. Network spilled Generation of deceitful data. Resource overload. Leak encryption keys and critical information (Jadidoleslamy 2014; Fatema and Brad 2013)	False identity, Misconduct, Deceitful routing and collision (Jadidoleslamy 2014; Fatema and Brad 2013)	Secure authentication, Secure routing, Secure Identity, Limited MAC rate, Small packet frames (Jadidoleslamy 2014; Fatema and Brad 2013)	False identity, Misconduct. Deceitful routing and collision (Taylor et al. 2014)	Use of symmetric key, a keying system that provides both forward and backward key secrecy (Taylor et al. 2014)
Wormholes	False routes, Overused Routing race conditions. Change in network topology. Collapse of path detection protocol. Packet destruction (Kaur and Singh 2014; Jadidoleslamy 2014)	False routing information. Techniques of Packet restrains (Kaur and Singh 2014; Jadidoleslamy 2014)	Multidimensional scaling algorithm, DAWWSEN protocol, Border control protocol, Graphical position system. Ultrasound, Sync in global clock. Authenticated encryption of link layer. Global shared key (Kaur and Singh 2014; Jadidoleslamy 2014)		Gianluca model for Wormhole attack and defense mechanism for cyber-physical systems (Dini and Tiloca 2014)
Unfairness	Efficiency reduction Demand for channel access. Limit in channel access capacity (Mohammadi and Jadidoleslamy 2014; Jadidoleslamy 2014)	Misbehavior (Jadidoleslamy 2014)	Small frames usage (Jadidoleslamy 2014)	Statistical analysis (Gill 2002)	Small frames usage (Gill 2002)
De-synchronization	Disruptive communication Drain of resources (Jadidoleslamy 2014; Mohammadi and	De-established connections (Jadidoleslamy 2014; Mohammadi and Jadidoleslamy 2011a; Sabeel et al. 2013)	Strong authentication mechanisms Time synchronization (Jadidoleslamy 2014; Mohammadi and	De-established connections between the nodes (Gill 2002	Strong Authentication (Gill 2002)

(continued)

Table 4.1 (continued)

Attack	Effects	WSN		CPS	
		Detection methods	Defensive methods	Detection methods	Defensive methods
	Jadidoleslamy 2011a; Sabeel et al. 2013		Jadidoleslamy 2011a; Sabeel et al. 2013		SCADA Security Device(SSD) to facilitate encrypted and authenticated communication with the master (Aniket 2009)
DoS attack	interfere with hosts as they access the local network (Jadidoleslamy 2014; Mohammadi and Jadidoleslamy 2011a)	Destroys networks links (Jadidoleslamy 2014; Mohammadi and Jadidoleslamy 2011a)	All of the above	Behavior, behavior-specification, knowledge (Aniket 2009)	
Internal attacks in data link layer					
Acknowledge spoofing	Falsifying data, packet loss, steer loops and adjust its length, broadcast blunder messages, modify and replay data tracked data (Mohammadi and Jadidoleslamy 2011a; Jadidoleslamy 2014)	Misbehavior	Validate, encrypt link layer and global shared key techniques with a new route (Mohammadi and Jadidoleslamy 2011a; Jadidoleslamy 2014; Singh et al. 2014)	Behavior of receiving data based on time (Park et al. 2010)	IDS, identity validation, strengthen the key negotiation process (Alcaraz and Lopez 2015)
External attacks in network layer					
Eavesdropping	Critical data extraction, exposed privacy (Jadidoleslamy 2014)	Statical methods (Jadidoleslamy 2014)	Control of access. Distributed nature of processing; Strong encryption (Jadidoleslamy 2014; Kaur and Singh 2014)	Behavior, behavior-specification, knowledge (Da Silva et al. 2015)	Mechanism uses the facilities provided by SDN to aid SCADA networks in the defense against unauthorized interception of flows by dispersing traffic across multiple paths (Da Silva et al. 2015)
Node subversion	Dishonesty, WSN down-ridden Exploitation of Resources (Kaur and Singh 2014)	No ACK from infected sensor (Kaur and Singh 2014)	Disguising sensors. Proper protocols. Restriction access, Data privacy (Kaur and Singh 2014)	Physical anomalies (Mcevoy and Wolthusen 2011)	An IP traced back protocol use to observe the packet behavior to locate and counter subverted nodes (Mcevoy and Wolthusen 2011)
Flooding	Exhaustion of resources			Formatted messages (Lopez 2012)	Use of encryption mechanisms introduces delays in

(continued)

Table 4.1 (continued)

Attack	Effects	WSN		CPS	
		Detection methods	Defensive methods	Detection methods	Defensive methods
	Downgraded traffic in WSN (Sun et al. 2014b; Virmani et al. 2014)	Complete network slows down (Sun et al. 2014b; Virmani et al. 2014)	Bidirectional authentication (Sun et al. 2014b; Virmani et al. 2014)		communication channels, signatures/verifications, key Management, TCP/IP countermeasures (Lopez 2012)
Spoofing	Network splitting, Overuse of Resources, Lifetime reduction in Network, Shedding of routing data (Sun et al. 2014b, Virmani et al. 2014)	Secure ARP Protocol Kernel-Based Patches, Passively Static MAC (Sun et al. 2014b, Virmani et al. 2014)	Encryption with MAC using different path for resending messages (Sun et al. 2014b, Virmani et al. 2014)	Behavior, behavior-specification, knowledge (Zhang et al. 2013b)	IDS: Premaratne Technique (Premaratne et al. 2010), cross-layer mechanism (Zhang et al. 2013b)
Wormholes	False corrupted routes, Overused Routing race conditions. Change in network topology, Collapse of path detection protocol, Packet destruction (Sun et al. 2014b; Virmani et al. 2014)	False routing information, Packet restrains (Sun et al. 2014b; Virmani et al. 2014)	Multidimensional scaling algorithm, DAWWSEN protocol, BCP, GPS, Ultrasound, Global clock, directional antennas, Global shared key (Sun et al. 2014b; Virmani et al. 2014)	Formatted messages (Kim 2012)	These attacks can be mitigated by establishing an isolation policy of malicious nodes using a specific threshold (Alcaraz and Lopez 2015)
DoS attacks	pound a target network with more data than it can handle (Sun et al. 2014b; Virmani et al. 2014)	False routing information, Packet restrains (Sun et al. 2014b; Virmani et al. 2014)	Multidimensional scaling algorithm, DAWWSEN protocol, BCP, GPS, Ultrasound, Global clock, directional antennas, Global shared key (Sun et al. 2014b; Virmani et al. 2014)	Behavior, behavior-specification, knowledge (Kang 2014)	IndusCAP-Gate system, automatically generates whitelists for traffic analysis and performs multiple filtering based on whitelists for blocking unauthorized access from external networks (Kang et al. 2014b)

Internal attacks in network layer

Misdirection	Corrupt routing tables, Overused Resources (Sun et al. 2014b)	Predicting delay and throughput (Sun et al. 2014b)	Hierarchical routing mechanism (Sun et al. 2014b)	Formatted messages (Vuković et al. 2011)	Security metrics to quantify the stealthy attacks (Vuković et al. 2011)

(continued)

Table 4.1 (continued)

Attack	Effects	WSN		CPS	
		Detection methods	Defensive methods	Detection methods	Defensive methods
Rushing	Discarding genuine requests Inability to discover any useful routes (Singh et al. 2014)	Inability to discover route more than two hops (Singh et al. 2014)	Rushing Attack Prevention (RAP) (Singh et al. 2014)	Formatted messages (Vuković et al. 2011)	An efficient certificate less signature scheme named McCLS based on the bilinear Diffie–Hellman assumption in the random oracle model (Xu et al. 2008)
Homing	Active attack on vital resources Extract critical network information, Threaten on data privacy (Singh et al. 2014)	Statistical methods (Singh et al. 2014)	Encryption (Singh et al. 2014)	Formatted messages (Mahan et al. 2011)	A firewall rule or access control rule should be created to specifically allow traffic from the Plant Data Historian to be exchanged with the SCADA or control system over specified ports at specified rates (Mahan et al. 2011)
Selective forwarding	Network disturbances with future attacks (Singh et al. 2014; Yu et al. 2012)	Statistical methods (Singh et al. 2014; Yu et al. 2012)	Network monitoring with source routing (Singh et al. 2014; Yu et al. 2012)	Formatted messages (Alcaraz and Lopez 2015)	Mitigation possible through dynamically select the next hop from a set of candidates working on the supposition that this set does not have any compromised nodes (Alcaraz and Lopez 2015)
Sybil attack	Causes network ineffectiveness in data integrity and Accessibility (Yu et al. 2012; Kaur and Singh 2014)	Disregard distrusted nodes through matching rules of trust ranking to counterfeit corresponding nodes in order to boost its own reputation. This will also help to reduce the reputation of others (Yu et al. 2012)	Physical shield of nodes (Yu et al. 2012; Kaur and Singh 2014)	Maintaining low overhead and delay of signals. (P2DAP) (Alcaraz and Lopez 2015)	Strengthen the key negotiation process. Additionally validate the identities of the nodes (Alcaraz and Lopez 2015)
Black hole	Suppress of broadcast over false shortest path with hub as black hole for data packets to disrupts network routing table with high packet loss rate	Disregard distrusted nodes through matching rules of trust ranking (Yu et al. 2012)	Surveillance, Multipath Routing, Decentralized IDS, Sensor network automated intrusion detection system (SNAIDS)	Traffic drain (Xu et al. 2008)	An efficient certificate less signature scheme named McCLS based on the bilinear Diffie–Hellman assumption in the random oracle model (Xu et al. 2008)

(continued)

Table 4.1 (continued)

Attack	Effects	WSN		CPS	
		Detection methods	Defensive methods	Detection methods	Defensive methods
Spoofing	Network split, Overrun of Resources, Lifetime reduction in Network, Shedding of routing data (Singh et al. 2014; Kaur and Singh 2014)	Secure ARP protocol, Kernel-based patches (Singh et al. 2014; Kaur and Singh 2014)	Encryption, Authentication with MAC Multipath message resend (Singh et al. 2014; Kaur and Singh 2014)	Behavior, behavior-specification, knowledge (Yang et el. 2014)	a SCADA-specific IDS is to identify external malicious attacks and internal unintended misuse. Access control whitelists, protocol-based whitelists, and behavior-based rules (Yang et el. 2014)
ACK spoofing	Selective forward attack Packet fraud (Singh et al. 2014; Kaur and Singh 2014)	Secure ARP protocol with kernel-based patches to detecting passively (Singh et al. 2014; Kaur and Singh 2014)	Sniffer ARP, MAC-ARP header, traffic filter, spoof detector, spoof alarm (Singh et al. 2014; Kaur and Singh 2014)	Behavior, behavior-specification, knowledge (Zhang et al. 2013b)	Cross-layer defense mechanism to combat spoofing in smart grid (Zhang et al. 2013b)
Hello flood	Flooding whole network with false HELLO packets from neighbors (Virmani et al. 2014)	Bidirectional link (Virmani et al. 2014)	Client puzzle (Virmani et al. 2014)	Proactive–reactive jamming (Kim 2012; Da Silva et al. 2015)	Rate limiting commands should be given to control devices to restrict and limit data flooding (Kim 2012)
Internet smurfing	Complete stud down of victim's computer (Virmani et al. 2014, Singh et al. 2014)	Network link flooded with useless data (Virmani et al. 2014, Singh et al. 2014)	Attacker's node is made to sleep Deactivate IP broadcasting at the network router (Virmani et al. 2014, Singh et al. 2014)	Proactive–reactive jamming (Kang et al. 2014b)	IP packet filtering, network intrusion detection (Kang et al. 2014b)
Flooding	Exhaustion of resources Retarded availability Downgraded traffic in WSN (Singh et al. 2014)	Complete network slows down (Virmani et al. 2014)	Bidirectional authentication (Virmani et al. 2014)	Proactive–reactive jamming (Kang et al. 2014b)	IP packet filtering, network intrusion detection (Kang et al. 2014b)
Gray hole	Undetected disruption in the network (Singh et al. 2014; Virmani et al. 2014)	Check time computed b/w RREQ and RNPS of neighbor dynamic routing (Singh et al. 2014; Virmani et al. 2014)	Multipath routing checkpoint-based scheme, multi-hop acknowledgement scheme (Singh et al. 2014; Virmani et al. 2014)	Drop packets selectively rather than all of them. Behavior-specification (Barbosa and Pras 2010)	IDS: barbosa technique (Barbosa and Pras 2010)

(continued)

Table 4.1 (continued)

Attack	Effects	WSN		CPS	
		Detection methods	Defensive methods	Detection methods	Defensive methods
Gratuitous detour	Limiting resources looping erratic routs, network breakout, misdirection (Jadidoleslamy 2014)	Deterioration in network performance (Jadidoleslamy 2014)	Pair-wise authentication, network layer authentication, central certificate authority, adopt validation techniques (Jadidoleslamy 2014)	Proactive–reactive jamming (Vijayalakshmi and Rabara 2011)	Dominant pruning method. The nodes in both lists periodically handshake with each other to prune the nonexistent nodes (S. Vijayalakshmi and Rabara 2011)
External attacks in transport layer					
Flooding	Exhaustion of resources Retarded availability/traffic (Sun et al. 2014b; Virmani et al. 2014)	Complete network slows down (Sun et al. 2014b; Virmani et al. 2014)	Identity verification protocol, Bidirectional authentication, Route Error (RERR) message (Fatema and Brad 2013)	Proactive–reactive jamming (Jin et al. 2011)	Crypto-based solutions to establish strong authentication, puzzle-based identification techniques, DNP3 Secure authentication, DNPsec (Jin et al. 2011)
De-synchronization	Disruptive network, going out the SYNC, resource collapse (Sun et al. 2014b)	Delay and warping in performance (Sun et al. 2014b)	Dual path check (Sun et al. 2014b)	Distortion of measurements (Vijayalakshmi and Rabara 2011)	Linear state estimation of local areas observed with the PMU measurements (Kolosok et al. 2015)
Internal attacks in application layer					
Non-repudiation	Launch of selective forward attack, Packet fraud (Sun et al. 2014b)	Presence of false, deceptive log files (Sun et al. 2014b)	Sensor node identity; detection mechanism (Sun et al. 2014b)	Presence of false, deceptive log files (Pidikiti et al. 2013).	Eradicated by the use of operator authentication. Each possesses a unique authentication credentials (Pidikiti et al. 2013)
Data aggregation distortion	Incorrect environ monitoring Disrupted aggregation of data Cross-layer attacks (Ozdemir and Xiao 2008)	Delay and warping in performance (Ozdemir and Xiao 2008)	Bidirectional authentication (Ozdemir and Xiao 2008)	Delay and warping in performance Behavior + knowledge (Shin et al. 2010)	A hierarchical two-level clustering approach can be used to develop a balance between security and efficiency (Shin et al. 2010)
SQL injection	Modify database, admin operations, commands issued to O.S (Rietta 2006)	Alien vault USM network IDS, Host IDS (Rietta 2006)	Data base intrusion detection System based on anomalies, signatures or honey tokens (Rietta 2006)	Alien vault USM network IDS, Host IDS (Rietta 2006)	Data base intrusion detection system based on anomalies, signatures, or honey tokens (Rietta 2006)

(continued)

Table 4.1 (continued)

Attack	Effects	WSN		CPS	
		Detection methods	Defensive methods	Detection methods	Defensive methods
Software tampering	Misused binary patch up Code replacement (A S Sastry et al. 2013; Xing et al. 2010a, b)	Software checkup (Sastry et al. 2013; Xing et al. 2010a, b)	Anti-tamper software, malware scanners and antivirus applications (Sastry et al., 2013; Xing et al. 2010a, b)	Behavior, behavior-specification, knowledge (Kim 2012)	Enforcing a strong encryption mechanism, Simple malicious code detection, isolation (Kim 2012)
Dictionary attacks	Tap encrypted message or document (Wang and Wang 2013)	Failed password trial (Wang and Wang 2013)	Limiting number of attempts, securing accounts after unsuccessful logins (Wang and Wang 2013)	Formatted messages (Bartman 2015)	Strong password policy, account lockout (Bartman 2015)
Cookie replay	Network Masquerade (Liu et al. 2015)	Data conflict caused by replay, message travel via untrusted mediators Masquerade (Liu et al. 2015)	Synchronization, session tokens, time stamping, one-time passwords Masquerade (Liu et al. 2015)	Behavior, behavior-specification (Pidikiti et al. 2013)	Time stamping techniques in the data transmission protocols (Pidikiti et al. 2013)

the mechanism to defend against P-DoS attack on cyber-physical systems by finding the packet receiving rate and also by predetermined packet arrival threshold.

4.4.2 External and Internal Attacks in Data Link Layer

Similarly, the data link layer of CPS and WSN is also exposed to various external and internal attacks, such as collision, resource exhaustion, traffic manipulation, eavesdropping, impersonation, wormholes, unfairness, de-synchronization, and DoS attack. Sun et al. (2014a) reviews the attacks and security protocols of wireless sensor protocols and provides the basic knowledge and security requirements for wireless sensor network. The study revealed that interference and corruption of packet data can be controlled through sending multiple versions of the same signal are transmitted at different time instants (time diversity) and with error correction codes, respectively (Sun et al. 2014a). Similarly, Gill (2002) also advised the utilization of error correction codes for corrupted data in cyber-physical system. The study provides the theoretical overview of security aspect in network embedded systems. Mohammadi and Jadidoleslamy (2011a) and Fatema and Brad (2013) studied the misbehavior of sensor nodes due to repeated collisions of data. The exhausted nodes due to repeated collision can be controlled through multiple techniques that are restriction of or limitation of data rate, change of modulation technique to time division multiplexing and also through identity protection of nodes in the sensor network (Fatema and Brad 2013). The study is structured to provide the countermeasures but with some limitations as the provided measure has not been practically implemented or tested on any test bed. Aniket (2009) ensured that real-time monitoring mechanism with correlation of event logs helps to rectify the complications caused by exhaustion of resources. The study can be acknowledged with its results as it includes the design and implementation of device security of cyber-physical systems. The authors also investigated the effects of impersonation in data link layer of wireless sensor network and utilized the sheltered routing method with small packet frames to counter the adversity of deceitful data on network. Taylor et al. (2014) utilized symmetric keys for back and forth communication in their model. The suggested model is compatible with existing SCADA network and can be integrated partially or fully in any SCADA network. Mohammadi and Jadidoleslamy (2011a) studied the multidimensional DAWWSEN protocol to counter the wormhole attack in wireless sensor network. Dini and Tiloca (2014) studied a framework to counter the attacks of wormholes in cyber-physical systems. The framework has been tested through simulation to identify the intensity and grades attacks for a counter defense mechanism. While Mohammadi and Jadidoleslamy (2011a) studied that strong authentication and time synchronization mechanisms at data link layer which can be utilized as a defense mechanism for denial of service (DoS) and de-synchronization attacks. Gill (2002) identified that de-synchronization attack which can be countered through strong authentication. Aniket (2009) developed a SCADA security device to facilitate the encrypted and authentic communication in cyber-physical system, although the data link layer is also vulnerable to internal acknowledge spoofing attack. This was evident with study (Mohammadi and Jadidoleslamy 2011a; Singh et al. 2014) which identified that utilization of encrypted link layer with global shared keys helped to counter the effects of acknowledge spoofing. Alcaraz and Lopez (2015) studied that the acknowledgment spoofing can be strengthened by key negotiation process with identity validation of nodes. Kaur and Singh (2014) studied the denial of service attack on data that can be countered through physical shield of nodes with regular change of keys.

4.4.3 External Attacks in Network Layer

Although the network layer of CPS and WSN is also susceptible to various external attacks, such as eavesdropping, node subversion, flooding, spoofing, wormholes, and DoS attacks. Jadidoleslamy (2014) and Kaur and Singh (2014) identified that strong encryption techniques and distributed nature of processing can address the issue of privacy in wireless sensor networks. Da Silva et al. (2015) identified that privacy of data can be obtained by dispersing the traffic over the multiple paths. This work is based on experimental study keeping in view of existing smart grid model comprising anti-eavesdropping mechanism for the security of private data in cyber-physical system. Kaur and Singh (2014) identified the solution for node subversion attack in wireless sensor network. The study revealed that proper protocols, data privacy, and restricted access on nodes can be utilized as a protective measure for node subversion attacks. Mcevoy and Wolthusen (2011) presented a protocol which can be used to observe the packet behavior and counter the subverted nodes. The study also suggested a probabilistic model which could be cost-effective in defeating node-based attacks. The model was evaluated mathematically and can be taken as a significant approach toward a low computational cost model. Sun et al. (2014a) studied the exhaustion of resources due to flooding in wireless sensor network whose adverse effects is complete network slow down. Sun et al. (2014a) identified a mechanism based on bidirectional authentication of packets can be vital to counter the adverse network slow down. On a contrary, Javier Lopez and Wolthusen (2012) identified that the use of encryption techniques, key management techniques and delays in communication channel of cyber-physical system can confront the adverse effects of flooding. Virmani et al. (2014), on the other hand, suggested that secure ARP and passively static MAC could control the spoofing in the network layer of wireless sensor network. Zhang et al. (2013a) identified a cross-layer mechanism as a better approach to counter the spoofing in cyber-physical system and presents a new McCLS scheme with security protocols and provided a comparison on Qualnet-based software with AODV routing protocol without having any protection mechanism. Sun et al. (2014a) studied a multidimensional scaling algorithm along with global shared key to counter the effect of wormholes. Alcaraz and Lopez (2015) studied that wormhole attack can be countered through an isolation policy of malicious nodes. The study provided an analysis of Zigbee PRO, WirelessHART, and ISA100.11a keeping in view the security vulnerabilities, threats, and available security measures in these security standards. The study also provided detailed security guideline for attacks on critical cyber-physical systems. Kang et al. (2014a) studied the application layer attacks and network layer attacks and proposed an Indus CAP-Gate system that blocks the malicious activity while accessing from the external network. The proposed intrusion detection/prevention system required regular signature updates for its compatibility, but due to limited access to the external network in SCADA, the system will be susceptible to external threats. Therefore, the researchers need to address the issue with security schemes to enhance the SCADA security at the network layer.

4.4.4 Internal Attacks in Network Layer

Accordingly, the network layer of wireless sensor network and cyber-physical system is also susceptible to numerous attacks, such as misdirection, rushing homing selective forwarding, Sybil attack, sinkhole, black hole, spoofing, acknowledgment spoofing, hello flood, smurfing, flooding, gray hole, and gratuitous detour. Table 4.1 illustrates the effects of attack on network layer of WSN and CPS with their detection and defensive mechanism. Sun et al. (2014a) identified that egress filtering, authentication, hierarchical monitoring mechanism can address the misdirection attack in cyber-physical system. Vukovic et al. (2011) studied security metrics to quantify the misdirection attack. The study provides the efficient algorithms with their numerical output to analyze the attack for single path routing and also for multipath routing. The results proved that intensity of attacks reduced by 50%, if attack cost increased by $\varGamma m = 2$ and $\varGamma m = 3$ (Vukovic et al. 2011). The mathematical model was tested and validated through numerical analysis but has not implemented practically. Singh et al. (2014) studied the network rushing, homing and selective forwarding attacks and provides the countermeasures such as attack prevention technique, encryption, and network monitoring with source routing for the abovementioned attacks. The study provides an overview of the different techniques to counter the effect of cybersecurity attacks on layers of wireless sensor network but has not cyber-physical systems. Xu et al. (2008) provides a certificate less signature scheme to counter the internal attack of rushing on network layer of wireless sensor network. The study has provided the theoretical analysis with other CLS protocol but has not been evaluated for the same parameters designated for McCLS scheme. Mahan (2011) specified an access control rule to counter the effect of homing attack on SCADA network. The study is based on the technical recommendations presented for department of energy of United States. The emphasis of study was on security and technical guidelines for installations of SCADA system in energy production zones. Alcaraz and Lopez (2015) believes that mitigation of selective forwarding is possible by selection of nodes dynamically from the set of uncompromised nodes. Yu et al. (2012) studied Sybil attack and sinkhole attack in wireless sensor network identified a trust-based mechanism to counter the impact of these attacks on wireless sensor network. The study revealed that the attack could either be mitigating with trust or the attack could violate the trust/reputation system itself. The study provides detailed overview of available trust mechanism for attacks and countermeasures. The study was not unique by the means of identify any new approach. Alcaraz and Lopez (2015) studied that the sinkhole attack and Sybil attack can be mitigated by adopting the isolation policy for malicious node and by strengthen the key negotiation process, respectively, in cyber-physical system. The study provides the review of available security schemes for high intensive cybersecurity but lack in uniqueness to identify any new security mechanism. Yang et al. (2014) identified the spoofing in cyber-physical system can be protected by protocol-based whitelist or either by behavior-based whitelist. The study proposed a SCADA-specific intrusion detection mechanism to identify the external and internal attack based on a protocol which can distinguish the malicious activity of compromised nodes. The peculiarity of the study is that, it has been simulated on a SCADA-specific test bed. Whereas Kaur and Singh (2014), identified that ACK spoofing can be controlled through sniffing the ARP MAC, ARP header and spoof detector or spoof filter. Zhang et al. (2013a) identified a cross-layer defense mechanism to combat the spoofing in smart grid system. The study suggested a cross-layer defense mechanism which has uniqueness of providing a new approach toward security of cyber-physical system security, and also has the peculiarity that the system has also been simulated for numerous attacks on cyber-physical systems. Virmani et al. (2014) studied an identity verification protocol as one of the defenses against for hello flood attack in network layer of wireless sensor network. Kim (2012) opinionated that the use of rate-limiting commands through control devices in cyber-physical system through which data limiting and flooding of devices can be countered in case of hello flood attack. The study has no hardware legacy and discusses the security mechanism available for 6LoWPAN in conjunction with IP security with advanced encryption schemes for cyber-physical systems. The work has no uniqueness as it is only discussing the available security mechanisms. Virmani et al. (2014) studied Internet smurfing, flooding, and gray hole attacks in wireless sensor network, where deactivating the IP broadcasting at network router, bidirectional authentication and multi-hop acknowledgement scheme, respectively, identified as counter mechanism for abovementioned attacks in network layer of wireless sensor network. The study is based on available literature in regard to various attacks on wireless sensor network. The study has no uniqueness in terms of any simulation or testing the attacks on any platform. Kang et al. (2014a) identified IP filtering and intrusion detection technique as a viable approach toward the flooding attack in cyber-physical system. The research has a peculiarity that industrial intrusion detection system is developed and being studied for different network layer and application layer protocol attacks. Barbosa and Pras (2010) developed a behavior-based intrusion detection system for anomalies detection in smart cyber-physical system. The study has hardware legacy and leading toward the vulnerability assessment of SCADA networks for threats to cyber-physical systems in real world. Jadidoleslamy (2014) identified the use of central certificate authority in wireless sensor network and pair to pair authentication with network layer authentication, the effects of gratuitous detour could be countered. The study provides the review of attacks and their defenses with their impact on wireless sensor network. The peculiarity of the study makes it different from other studies in the field as different dimensions of wireless sensor network security have been discussed in detailed. Vijayalakshmi and Rabara (2011) provided a dominant pruning method which is the most viable approach to counter the effect of gratuitous detour in cyber-physical system. The approach follows the handshaking mechanism to prune the non-existent nodes on the network on the lists updated periodically.

4.4.5 External Attacks in Transport Layer

The transport layer of wireless sensor network and cyber-physical system is also susceptible to attacks. Table 4.1 illustrates the effects of external attacks on transport of CPS and WSN with their detection and defensive mechanism. Fatema and Brad (2013) identified that exhaustion of nodes due to flooding can be counter with bidirectional authentication of a link with encrypted echo back mechanism. Jin et al. (2011) studied the event-based flooding attack in cyber-physical system. The study analyzed the flooding attack and its countermeasures on DNP-3 protocol SCADA, where the study recommends crypto-based solutions for authentication purpose of exchanging packet data in DNP-3 protocol. In addition to that, the study also recommends the puzzle-based identification techniques and DNP secure authentication mechanism for the purpose of authorization in SCADA systems. Sun et al. (2014a) briefly studied the de-synchronization attacks in transport layer of wireless sensor network, where Sun et al. identified that through process of authentication between the layers, the effects attacks can be countered. The study analyzed attacks on layers of wireless sensor network and also recommended some countermeasures for these. Kolosok et al. (2015) studied the linear state estimation algorithm to counter effect of de-synchronization in cyber-physical system. The outcome of the study based on the simulation for power-based SCADA networks. The study has a uniqueness that it provides a way toward the direction of state estimation method for PMU measurement in power-based SCADA networks.

4.4.6 Internal Attacks in Application Layer

However, the application layer of cyber-physical system and wireless sensor network is also susceptible to some internal attacks, such as non-repudiation, data aggregation distortion, SQL injection, software tampering, dictionary attack, and cookie replay. Table 4.1 illustrates the effects of internal attacks on the application layer of WSN and CPS with their detection and defensive mechanism. Sun et al. (2014a) discussed non-repudiation attack as selfish attack as in this attack the nodes deny to participate with other nodes. Sun et al. (2014a) identified a proper detection mechanism for the sensor node identification a viable solution to counter the effect of non-repudiation in wireless sensor networks. Pidikiti et al. (2013) identified a distinct approach in which compromised nodes due to non-repudiation can be removed by the authorized operators. Apart from the discussion about the security mechanism in SCADA, this study also suggested an experimental model for IEC 60870-5-104 protocol with a security hardener that is actually a single board computer for the implementation of security mechanism in application layer of SCADA. Ozdemir and Xiao (2008) provides a bidirectional authentication scheme to prevent wireless sensor network from data aggregation and distortion. On a contrary, a two-level hierarchical approach is followed by Shin et al. (2010) for data aggregation and distortion. The strength of the study is based on the numerical analysis of the result whereas; however, there are few shortcomings in terms of non-implementation on hardware. Rietta (2006) identified the approach to counter the effects of SQL injection through database intrusion detection system based on anomalies, signatures, or honey tokens. The study has uniqueness that the adopted mechanism has been tested through static analysis and also by dynamic monitoring of the model. Where in static analysis, the IDS automatically build the queries, whereas, in dynamic part, the run time monitoring of queries is possible and can be compared with statistically build model. In addition to that, a defensive mechanism is utilized to create a random set of instruction to build events which are not familiar to the attacker. Xing et al. (2010a, b) and Sastry et al. (2013) identified that software tempering can be countered through malware scanners and antivirus applications. Kim (2012) accentuated a strong encryption mechanism for a counter measurement against software tampering. Wang and Wang (2013) studied that, by limiting the number of attempt on the user account on application layer, the wireless sensor network can be cured. Bartman (2015) accentuated the strong password policy and account locked out procedure to prevent the system from dictionary attacks. The work provides an introduction to the cyberattack mitigation techniques for securing critical industrial system and also intrusion detection techniques to secure the system from cyberattacks. The article has not provided any specific data about its simulated results and its implementation to clear about its viability. Liu et al. (2015) accentuated to utilize synchronization session token and time stamping technique to counter the cookie replay attack in application layer of wireless sensor network. Pidikiti et al. (2013) also advised to use the time stamping technique in data transmission protocol of cyber-physical system. The study provides a hardware legacy to counter the effects of cookie replay in cyber-physical system. In addition to that, authentication security model accompanied AES (128 bits) algorithm has also been introduced to enhance the viability of the model.

4.5   Conclusion

The recent work integrates the available threats and their available defensive mechanism for both cyber-physical systems and wireless sensor network. The cyber-physical system and wireless system are having some potential applications. These applications are of strategic nature and require proper security defense mechanism against these challenges. The layered structure, wireless and shared nature of communication, of both wireless sensor network and cyber-physical system makes it more susceptible to intruders. The study provides layer-to-layer analysis of security mechanism available for both wireless and cyber-physical system. In addition, the emphasis of study was on comparison between available defense mechanism for both cyber-physical system and wireless sensor network. In addition to that, a layer-wise table (Table 4.1) is also presented, which summarizes the nature of attack and their counter defensive mechanisms.

References

Alcaraz, C., & Lopez, J. (2015). A security analysis for wireless sensor mesh networks in highly critical systems. *IEEE Transactions on Systems, Man, and Cybernetics.*

Ali, S., & Anwar, R. W. (2012). Trust based secure cyber physical systems. In *Workshop Proceedings: Trustworthy Cyber-Physical Systems, Newcastle University: Computing Science, 2012.*

Ali, S., Qaisar, S. B., Saeed, H., Khan, M. F., Naeem, M., & Anpalagan, A. (2015). Network challenges for cyber physical systems with tiny wireless devices: A case study on reliable pipeline condition monitoring. *Sensors, 15,* 7172–7205.

Aloul, F., Al-Ali, A., Al-Dalky, R., Al-Mardini, M., & El-Hajj, W. (2012). Smart grid security: Threats, vulnerabilities and solutions. *International Journal of Smart Grid and Clean Energy, 1,* 1–6.

Aniket, R. (2009). *SCADA security device; Design and implementation.* MS Masters, Wichita university.

Barbosa, R. R. R., & Pras, A. (2010). Intrusion detection in SCADA networks. In *Proceedings of the Mechanisms for Autonomous Management of Networks and Services, and 4th International Conference on Autonomous infrastructure, Management and Security* (pp. 163–166). Berlin.

Bartman, T. (2015). How to secure your SCADA system. *The Journal.* Rockwell Automation.

Cardenas, A. A., Amin, S., & Sastry, S. (2008). Secure control towards survivable cyber-physical systems. In *Distributed Computing Systems Workshops, 2008. ICDCS '08. 28th International Conference on.* Beijing: IEEE.

Da Silva, E. G., Knob, L. A. D., Wickboldt, J. A., Gaspary, L. P., Granville, L. Z., & Schaeffer-Filho, A. (2015). Capitalizing on SDN-based SCADA systems: An anti-eavesdropping case-study. In *IFIP/IEEE International Symposium on Integrated Network Management (IM).* Ottawa, Canada: IEEE.

Deng, J., Richard, H., & Mishra, S. (2004). Countermeasures against traffic analysis attacks in wireless sensor networks CU-CS-987-04. *Computer Science Technical Report.*

Dini, G., & Tiloca, M. (2014). A simulation tool for evaluating attack impact in cyber physical systems. In *Modelling and simulation for autonomous systems.* Springer.

Fatema, N., & Brad, R. (2013). Attacks and counter attacks on wireless sensor networks. *International Journal of Ad Hoc, Sensor & Ubiquitous Computing (IJASUC), 4.*

Gill, D. H. (2002). *New vista in CIP researh and development: Secure network embedded systems.* Leesburg, VA: NSF/OSTP.

Govindarasu, M., Hann, A., & Sauer, P. (2012). Cyber-physical systems security for smart grid. In *The future grid to enable sustainable energy systems.* PSERC Publication.

Han, G., Jiang, J., Shu, L., Niu, J., & Chao, H.-C. (2014). Management and applications of trust in wireless sensor networks: A survey. *Journal of Computer and System Sciences, 80,* 602–617.

Jadidoleslamy, H. (2014). A comprehensive comparison of attacks in wireless sensor networks. *International Journal of Computer Communications and Networks (IJCCN), 4.*

Javier Lopez, R. S., & Wolthusen, S. D. (2012). *Critical infrastructure protection.*

Jin, D., Nicol, D. M., & Yan, G. (2011). An event buffer flooding attack in DNP3 controlled SCADA system. In *Winter Simuation Conference, 2011* (pp. 2614–2626). Phoenix, AZ.

Kang, D.-H., Kim, B.-K., & Na, J.-C. (2014a). Cyber threats and defence approaches in SCADA systems. In *16th International Conference on Advanced Communication Technology.* Pyeongchang: IEEE.

Kang, D., Kim, B., Na, J., & Jhang, K. (2014b). Whitelists based multiple filtering techniques in SCADA sensor networks. *Journal of Applied Mathematics.*

Kaur, D., & Singh, P. (2014). Various OSI layer attacks and countermeasure to enhance the performance of WSNs during wormhole attack.

Kim, H. (2012). Security and vulnerability of SCADA systems over IP-based wireless sensor networks. *International Journal of Distributed Sensor Networks, 2012.*

Kim, K.-D., & Kumar, P. (2013). An overview and some challenges in cyber-physical systems. *Journal of the Indian Institute of Science, 93,* 341–352.

Kolosok, I., Korkina, E., & Gurina, L. (2015). Vulnerability analysis of the state estimation problem under cyber attacks on WAMS. In *International Conference on Problems of Critical Infrastructures.* Saint Petersburg.

Li, Z., & Gong, G. (2008). Survey on security in wireless sensor. *Special English Edition of Journal of KIISC, 18,* 233–248.

Lin, C.-Y., Zeadally, S., Chen, T.-S., & Chang, C.-Y. (2012). Enabling cyber physical systems with wireless sensor networking technologies. *International Journal of Distributed Sensor Networks, 2012.*

Liu, A. X., Kovacs, J. M., Huang, C. T., & Gouda, M. G. (2015). A secure cookie protocol. In *14th International Conference on Computer Communications and Networks, 2005 (ICCCN 2005)* (pp. 333–338). 17–19 Oct. 2005, San Diego, California, USA: IEEE.

Lopez, J., Roman, R., & Alcaraz, C. (2009). Analysis of security threats, requirements, technologies and standards in wireless sensor networks. In *Foundations of Security Analysis and Design V.* Springer.

Lu, T., Zhao, J., Zhao, L., Li, Y., & Zhang, X. (2014). Security objectives of cyber physical systems. In *7th International Conference on Security Technology (SecTech),* (pp. 30–33). IEEE.

Mahan, R. E, Burnette, J. R., Fluckiger, J. D., Goranson, C. A., Clements, S. L., Kirkham, H., Tew, C. (2011). *Secure data transfer guidance for industrial control and SCADA systems.* Washington: Pacific Northwest National Laboratory Richland.

Mahmood, M. A., Seah, W. K. & Welch, I. (2015). Reliability in wireless sensor networks: A survey and challenges ahead. *Computer Networks.*

Mao, X., Zhou, C., He, Y., Yang, Z., Tang, S., & Wang, W. (2011). Guest editorial: Special issue on wireless sensor networks, cyber-physical systems, and internet of things. *Tsinghua Science and Technology, 16,* 559–560.

McEvoy, T. R., & Wolthusen, S. D. (2011). Defeating network node subversion on SCADA systems using probabilistic packet observation. *International Journal of Critical Infrastructures, 9,* 32–51.

Mitchell, R., & Chen, I.-R. (2014). A survey of intrusion detection techniques for cyber-physical systems. *ACM Computing Surveys (CSUR), 46,* 55.

Mo, Y., Kim, T.-H., Brancik, K., Dickinson, D., Lee, H., Perrig, A., et al. (2012). Cyber–physical security of a smart grid infrastructure. *Proceedings of the IEEE, 100,* 195–209.

Mohammadi, S., & Jadidoleslamy, H. (2011a). A comparison of link layer attacks on wireless sensor networks. *arXiv preprint* arXiv:1103.5589.

Mohammadi, S., & Jadidoleslamy, H. (2011b). A comparison of physical attacks on wireless sensor networks. *International Journal of Peer to Peer Networks, 2,* 24–42.

Muhammad Farhan, R. S. (2014). Energy efficient clustering algorithms for wireless sensors network: A survey. In *International Conference on Computers and Emerging Technologies (ICCET-2014),* Khairpur, Pakistan.

Networks, T. (2016). *A modern approach to safeguarding your industrial control systems and assets.* Available: https://www.temperednetworks.com/wp-content/uploads/2015/05/Tempered Networks-Securing-Industrial-Control-Systems.pdf.

Newsome, J., Shi, E., Song, D., Perrig, A. (2004) The sybil attack in sensor networks: Analysis & defenses. In *Third International Symposium on Information Processing in Sensor Networks, 2004. IPSN 2004.* 6–27 April 2004 California, USA: IEEE.

Ozdemir, S., & Xiao, Y. (2008). Secure data aggregation in wireless sensor networks: A comprehensive overview. *Computer Networks, 53,* 2022–2037.

Pal, P., Schantz, R., Rohloff, K., & Loyall, J. (2009). Cyber-physical systems security-challenges and research ideas. In *Workshop on Future Directions in Cyber-physical Systems Security.*

Park, K., Lin, Y., Metsis, V., Le, Z., Makedon, F. (2010). Abnormal human behavioral pattern detection in assisted living environments. In *Proceedings of the 3rd International Conference on Pervasive Technologies Related to Assistive Environments.* ACM.

Pidikiti,D. S., Kalluri, R., Senthil Kumar, R. K., & Bindhumadhava, B. S. (2013). SCADA communication protocols: Vulnerabilities, attacks and possible mitigations. *CSIT, 1,* 135–141.

Premaratne, U. K., Samarabandu, J., Sidhu, T. S., & Beresh, R. (2010). An intrusion detection system for IEC61850 automated substations. *IEEE Transactions on Power Delivery, 25,* 2376–2383.

Rietta, F. S. (2006). Application layer intrusion detection for SQL injection. In *44th ACM South East Conference, 2006.* University Blvd, Melbourne, Florida, USA, FL 32901. ACM.

Sabeel, U., Maqbool, S., & Chandra, N. (2013). Categorized security threats in the wireless sensor networks: Countermeasures and security management schemes. *International Journal of Computer Applications, 64,* 19–28.

Saqib, A., Waseem, R. A., & Hussain, O. K. (2015). Cyber security for cyber-physical systems: A trust based approach. *71,* 144–152.

Sastry, A. S., Sulthana, S., & Vagdevi, S. (2013). Security threats in wireless sensor networks in each layer. *International Journal in Advanced Networking and Applications, 4,* 1657–1661.

Shafi, Q. (2012). Cyber physical systems security: A brief survey. In *ICCSA Workshops, 2012* (pp. 146–150).

Shin, S., Kwon, T., Jo, G.-Y., Park, Y., & Rhy, H. (2010). An experimental study of hierarchical intrusion detection for wireless industrial sensor networks. *IEEE Transactions on Industrial Informatics, 6,* 744–757.

Singh, H., Agrawal, M., Gour, N., & Hemrajani, N. (2014). A study on security threats and their countermeasures in sensor network routing. *Prevention, 3.*

Stelte, B., & Rodosek, G. D. (2013). Assuring trustworthiness of sensor data for cyber-physical systems. In *2013 IFIP/IEEE International Symposium on, Integrated Network Management (IM 2013),* (pp. 395–402). IEEE.

Sun, F. M., Zhao, Z., Du, L., & Chen, D. (2014a). A review of attacks and security protocols for wireless sensor networks. *Journal of Networks, 9,* 1103–1113.

Sun, F., Zhao, Z., Fang, Z., Du, L., Xu, Z., & Chen, D. (2014b). A review of attacks and security protocols for wireless sensor networks. *Journal of Networks, 9,* 1103–1113.

Taylor, C. R., Shue, C. A., & Paul, N. R. (2014). A deployable SCADA authentication technique for modern power grids. In *International Conference on Energy.* Cavtat: IEEE.

Vijayalakshmi, S., & Rabara, S. A. (2011). Grilling gratuitous detour in Adhoc network *International Journal of Distributed and Parallel Systems (IJDPS), 2.*

Virmani, D., Soni, A., Chandel, S., & Hemrajani, M. (2014). Routing attacks in wireless sensor networks: A survey. *arXiv preprint* arXiv:1407.3987.

Vuković, O., Sou, K. C., Dán, G., & Sandberg H. (2011). Network-layer protection schemes against stealth attacks on state estimators in power systems. In *IEEE International Conference on Smart Grid Communications (SmartGridComm)*. Brusseles: IEEE.

Wan, J., Chen, M., Xia, F., Di, L., & Zhou, K. (2013a). From machine-to-machine communications towards cyber-physical systems. *Computer Science and Information Systems, 10,* 1105–1128.

Wan, K., Man, K., & Hughes, D. (2010). Specification, analyzing challenges and approaches for cyber-physical systems (CPS). *Engineering Letters, 18,* 308.

Wan, J., Yan, H., Liu, Q., Zhou, K., Lu, R., & Li, D. (2013b). Enabling cyber–physical systems with machine–to–machine technologies. *International Journal of Ad Hoc and Ubiquitous Computing, 13,* 187–196.

Wang, C., Sohraby, K., Li, B., Daneshmand, M., & Hu, Y. (2006). A survey of transport protocols for wireless sensor networks. *IEEE Network, 20,* 34–40.

Wang, D., & Wang, P. (2013). Offline dictionary attack on password authentication schemes using smart cards. In *Proceedings 16th Information Security Conference (ISC 2013)*. Dallas, TX: Springer-Verlag.

Wood, A. D., Srinivasan, V., & Stankovic, J. A. (2009). Autonomous defenses for security attacks in pervasive CPS infrastructure. In *Proceedings DHS: S&T Workshop on Future Directions in Cyber-physical Systems Security*.

Wu, F.-J., Kao, Y.-F., & Tseng, Y.-C. (2011). From wireless sensor networks towards cyber physical systems. *Pervasive and Mobile Computing, 7,* 397–413.

Xia, F., Mukherjee, T., Zhang, Y., & Song, Y.-Q. (2011). Sensor networks for high-confidence cyber-physical systems. *International Journal of Distributed Sensor Networks, 2011.*

Xing, K., Srinivasan, S. S. R., Jose, M., Li, J., & Cheng, X. (2010a). Attacks and countermeasures in sensor networks: A survey. In *Network Security*. Springer.

Xing, K., Srinivasan, S. S. R., Rivera, M., Li, J., & Cheng, X. (2010b). *Attacks and countermeasures in sensor networks: A survey*. Hefei, Anhui, China: Springer Science and Business Media Cell.

Xu, Z., Liu, X., Zhang, G., He, W., Dai, G., & Shu, W. (2008). A certificateless signature scheme for mobile wireless cyber-physical systems. In *The 28th International Conference on Distributed Computing Systems Workshops*. Beijing: IEEE.

Yang, Y., McLaughlin, K., Sezer, S., Littler, T., Im, E. G., Pranggono, B., et al. (2014). Multiattribute SCADA-specific intrusion detection system for power networks. *IEEE Transactions on Power Delivery, 29,* 1092–1102.

Yu, Y., Li, K., Zhou, W., & Li, P. (2012). Trust mechanisms in wireless sensor networks: Attack analysis and countermeasures. *Journal of Network and Computer Application, 35,* 867–880.

Zhang, Z., Trinkle, M., Dimitrovski, A. D. (2013a). Combating time synchronization attack: A cross layer defense mechanism. In *IEEE/ACM International Conference on Cyber-Physical Systems, ICCPS, 2013,* (pp. 141–149). New York: IEEE/ACM.

Zhang, L., Wang, Q., & Tian, B. (2013b). Security threats and measures for the cyber-physical systems. *The Journal of China Universities of Posts and Telecommunications, 20,* 25–29.

Chapter 5
ICS/SCADA System Security for CPS

Cyber-physical systems (CPSs) are a various collection of information communication technology (ICT) and embedded microprocessors which are communicated to the physical world via sensors and actuators. Smart grids are one area of applications of CPS. The remote activities of the smart grid's CPSs are monitored and controlled by specialized computing system called industrial control systems (ICSs) or supervisory control and data acquisition (SCADA) systems (ICS/SCADA). Hence, it is crucial to keep ICS/SCADA system safe and secure to prevent any cyberattack causing a physical hazard to the smart grid which might affect the human life, national safety, or economy.

The ICS/SCADA system should be designed and implemented following the most practical security practices as defined by well-known standards, guidelines, best practices, and policies. In addition, there are a numerous number of security knowledge's resources suggesting various methods to protect ICS/SCADA systems. However, this overloaded and scattered information will create difficulties for the organization to grasp the full picture of the ICS/SCADA security issues and the protection requirements which might lead to wrong, incomplete, or weak decisions.

5.1 Introduction

Cyber-physical system is working as a bridge between the physical environment and the cyber world via little or a group of cooperated sensors and actuators to act with heterogeneous data flow which, at the end, will be controlled by an intelligent decision control systems (Wu et al. 2011). Also, the commercial of CPS's innovations will not be possible without the innovative technology of sensors, actuators, communication, the internet, and wireless revolution besides the lower cost of embedded electronic devices (Rajkumar et al. 2010). The CPS's commercial implementations can be utilized in monitoring and controlling the water supply, oil

© Springer International Publishing AG 2018
S. Ali et al., *Cyber Security for Cyber Physical Systems*, Studies in Computational Intelligence 768, https://doi.org/10.1007/978-3-319-75880-0_5

production, transport, telecommunication, and electricity power generation and transmission (Sanislav and Miclea 2012). Therefore, the operations of CPSs need to be safe to avoid any hazard to the human and the surrounding environment, by ensuring and maintaining the security properties confidentiality, integrity, and availability (CIA) (Aditya 2015). In addition, any cyberattacks targeting CPSs working for the critical infrastructure such as smart grid will have a great negative impact; it might go further of losing the human lives (Mo et al. 2012). These cyberattacks could be done through many ways such as eavesdropping, denial of service (DoS), and the wireless jamming (Ali et al. 2015). The diversity of devices and connection will create a great chance for more vulnerabilities for attacking the sensitive and private information of the surrounding physical environment (Amin et al. 2013).

With the diversity of security attacks in CPSs, the protection of their applications, networks, and servers is critical to keep them safeguarded from any illegal accesses and malicious attacks. This is especially important in critical mission infrastructures as smart grid, in which the security must be reliable and have a secure environment to run the routine operations (Kui et al. 2011). Moreover, it is obvious to protect and secure the CPS's devices specifically which are used in the nationwide infrastructure including the smart grid (Charalambos et al. 2015). To protect the smart grid, a dedicated security framework and policy should be adopted and utilized. This framework should provide a common understanding of the uniqueness of CPS's security which will enhance sharing knowledge and integrating the security research and support development of more security activities toward the CPS's applications such as smart grid (NIST 2016).

5.2 ICS/SCADA Systems Security

Nowadays, ICSs are used in national critical infrastructure (such as smart grid) to monitor, manage, and control various critical processes and physical functions, remotely and in a real-time manner (Alshami et al. 2008). ICS is a general term and includes many various types of control systems such as SCADA and distributed control systems (DCSs). In practice, the different media publications normally use a combination of ICS and SCADA words as a form of operational technology (OT) research topics (Darktrace 2015).

ICS/SCADA systems are meant to acquire data from various remote instruments such as pumps, transmitters, valves, etc. and control them remotely from a host's central SCADA software. Further, it is composed of computers, networks, and embedded devices which work together to monitor, manage, and control vital processes in various industrial and critical infrastructure areas such as oil and power management. This combination consists mainly of field instrumentation, programmable logic controllers (PLCs) and/or remote terminal unit (RTUs), communications networks, and ICS/SCADA host software (Schneider 2012).

Smart grid's field instruments are monitored and controlled remotely through ICS/SCADA control room (Coates et al. 2010). Also, the power's real-time data should be secured and available continuously to provide reliable and best power management (Gao et al. 2013). For that, smart grid should be well protected from such cyberattacks (Shuaib et al. 2015). Also, protecting such critical infrastructure is vital and challenging at the same time because a minimum downtime can cause many problems. Therefore, smart grid must carefully identify the security problems of ICS/SCADA systems, the possible attack vectors, evaluate and rank the different threats, and remediate the possible vulnerabilities (Khurana et al. 2010).

Additionally, understanding the security requirements of the ICS/SCADA systems requires to grasp enough knowledge about the ICS/SCADA architecture, vulnerabilities, and attacks, and it motivates attack categories, security challenges, objectives, and planning to implement the security program. The following sections will be going to highlight briefly all these concepts.

5.2.1 Architecture of ICS/SCADA

As shown in Fig. 5.1, a conceptual illustration of ICS/SCADA infrastructure has four main components, namely the ICS/SCADA control center, communication networks, PLCs and/or RTUs, and the field instruments (Schneider 2012).

ICS/SCADA control room consists of servers which include OPC and database, and the end user computers and HMI. OPC server is used as a software interface to allow Windows software to communicate with the industrial hardware instruments, like the concept of the object linking and embedding (OLE), but for process control (Galloway and Hancke 2013). Communication networks allow connectivity between SCADA control room, PLCs, RTUs, and the various field instruments. It can use numerous communication technologies and devices such as fiber, Ethernet, Wi-Fi, switch, routers, modems, satellite, etc. (Kuzlu et al. 2014).

PLCs are rugged specific industrial computers with predefined instructions stored in their memory to do certain tasks, and always observe the state of its connected input devices such as sensors, and act based upon custom programming codes to change the state of output devices such as actuators. Also, PLCs have

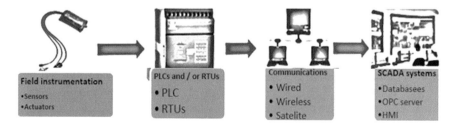

Fig. 5.1 Conceptual illustration of ICS/SCADA infrastructure developed

inputs/outputs to interact with field instruments such as sensors, actuators, valves, pumps, etc. and keep continuously observing the state of the various input devices and make logic-based decisions upon predefined custom instruction to control the state of output devices. Therefore, ICS/SCADA systems depend totally on PLCs to monitor and control the field instruments. In addition, RTU is a telemetry electronic device that connected to remote input devices such as sensors and receives input data from them and transfers it to the ICS/SCADA control room. PLCs and RTUs are becoming one of the core technologies used in smart grid environments (Moscatelli 2011).

ICS/SCADA system's field instruments consist of miscellaneous instruments and devices such as temperature sensor, pressure sensor, level sensor, flow meter, various valves, smart pumps, etc. Also, sensors are small electronic devices that sense the surrounding physical quantities measurements such as voltage, temperature, pressure, speed, etc. These sensors send the collected physical data to the ICS/SCADA controlling servers for further analyzing and monitoring. Actuators are devices that are responsible for making action to the surrounding or to the attached environment's components, and the action might be mechanical or electronic. Also, these devices will act upon receiving controlled orders from the ICS/SCADA control room. These instruments are critical analog input devices to the PLCs and modern RTUs to process the input data and make the logic-based decision accordingly (Ida 2014).

ICS/SCADA architecture in smart grid is almost similar with some unique features. For instance, in the smart grid, the utility's substations send the most real-time electrical data status to SCADA control room via PLCs or/and RTUs. Also, the modern substations are occupied with intelligent electronic device (IED) such as electronic circuit breakers and power monitors which cooperate with PLCs/RTUs to transmit the collected data to the substation's computer which concurrently will be transferred to the main centralized SCADA system (Thomas and McDonald 2015).

5.3   ICS/SCADA Vulnerabilities

Cyber-physical's embedded systems are the operation fundamental of the ICS/SCADA systems which become more vulnerable to exploitation; therefore, they should be well-protected and secured (Cárdenas et al. 2009). The vulnerability of the ICS/SCADA system is due to many factors such as the absence of real-time network scanning to detect suspicious activities, identify the threats, and react accordingly, the slowness for systematic and careful updating and patching, and lack of knowledge about the specification and capabilities of the old and new devices. Also, not having enough information about the network's traffic status is the core reason for the vulnerability because it stops the security specialists to know about any abnormal activities or potential threats to field instruments and ICS/SCADA system. Neglected, unskilled, and unsafe practices of authentication give chances to the attackers to gain access to these holes (Kim and Kumar 2013).

5.4 Attacking Methods Toward ICS/SCADA

An attacker targeting the ICS/SCADA systems are not in one trial but through a bulkiness of efforts and methods to gather the most applicable and adequate information to create a great negative cause. The type and complexity of attacking methods that are used are determined by how much the attacking purpose is vital, what is the impact level must be achieved to satisfy these purposes, and how much the ICS/SCADA system is secured. For instance, if an attacker has no harm objective, s/he might satisfy using DoS. However, s/he needs to go further to destruct the ICS/SCADA system by manipulating the operational process to achieve the ultimate damage and harms to the structure, equipment, data, and human. Then, a huge systematically attacking method will be used (Lee et al. 2015). Dell listed the most methods that are being used to attack the ICS/SCADA system in 2014 (Dell 2015).

5.5 Threats Categories Facing ICS/SCADA

The security of the ICS/SCADA operations is threatened by two main categories of intimidation including unintentional and purposeful threats. Unintentional (inadvertent) threats are mainly originated from inside the premises, and the primary sources are the human, devices, or the surrounding nature. The human factor including the employees, contractors, and/or business partners can be a source of threats in the form of neglect, carelessness, or lack of knowledge, so s/he might produce a threat toward the ICS/SCADA system without awareness or intention to do that. Also, the machine's failure, devices safety weakness, and equipment crashing are a great source of threats to the ICS/SCADA security (IEC 2016). Moreover, the natural disasters such as avalanches and landslides, earthquakes, sinkholes, volcanic eruptions, floods, tsunami, and blizzards can cause disruption to the ICS/SCADA security (Laing et al. 2012). Additionally, the ICS/SCADA system can be targeted by the meaning of purposeful attacks by annoyed employees, industrial espionage, sabotage, cyber hackers, viruses and worms, physical theft, and electronic terrorism (IEC 2016).

5.6   The Challenges to Protect ICS/SCADA

There are many security challenges facing the professionals to protect the ICS/ SCADA systems such as the old and basic architecture designs in which the security matter is not considered. Also, still the message transmitted internally and between remote connections using clear text form without adopting any encryption technologies. There are many applications and servers' operating systems and applications that have no regular patching scheme, even some firmware not updated at all. Another challenge is to secure the remote communication of the ICS/SCADA components which are geographically scattered such as sensors, actuators, RTUs, and PLC. In addition, there are several management, operational, and technical challenges to secure the ICS/SCADA systems, such as vulnerability tracking problem, standardization of devices and systems, downtime for maintenance, unsupported OS and applications, exposed to OT network to public networks, unable to pen-test in production real time, limited time for remediation once incidents occur, share accounts or no authentication within users, and secure the connection between information technology (IT) and OT (Babu et al. 2017).

5.7   ICS/SCADA Security Objectives

There are many objectives to protect the ICS/SCADA architectures such as keeping the ICS/SCADA systems functioning as much as possible during the most difficult conditions. This involves creating a redundancy for the most ICS/SCADA critical devices such as historian servers, switches, modems, and the field instrumentations. Also, during a failure, the device should not cause unnecessary traffic on the ICS/ SCADA networks or cause extra trouble elsewhere.

ICS/SCADA system should be designed to allow for systematic degradation such as transferring from "normal condition" with full automation to "emergency condition" with operators interfere making less automation to "full manual status" with no automation at all (NIST 2015). Another protection objective is to provide a systematic and practical way to detect the various security events and incidents. For instance, security specialists can safeguard the operations of ICS/SCADA if they can detect early the failure of devices, services exhausted resources such as memory, processing, and bandwidth that are being used (Ralston et al. 2007).

Furthermore, there are three traditional security objectives for securing the ICS/ SCADA systems which are availability, integrity, and confidentiality (AIC). Availability is on top of security objectives priority list which is focusing on keeping ICS/SCADA systems services offered continuously 24/7 when monitoring and controlling critical infrastructure or life-safety systems. Also, integrity comes the second order in the list by the ability to offer the operators/controllers the confident needed to fully trust the integrity of the various physical information that is received and to take the most suitable actions based on reading feedback or status from the various instruments and devices. In addition, confidentiality is not as important as the availability and integrity because the received physical information from sensors, PLCs and RTUs, are used and the transmitted is state-based and only valid for that specific time; then, it will be discarded after processing and storing them in the historian servers. For instance, the living time for physical data to be processed is very short and the period is between two processed physical data as short as milliseconds in contrast with the traditional IT data where credit card details are effective for many years (Homeland-Security 2009).

Park and Lee (2014) have conducted a research to find whether a new set of security objectives are required or will be similar to IT typical security objectives: CIA. The researchers compared security requirements of three international standards, namely ISO 27001, NIST SP 800-53, and IEC 61511 to check whether the ICS/SCADA safety is considered by the above standards. Also, these standards consider the safety as an important value in ICS/SCADA security equally with CIA. Also, they found that security objectives based on ISO 27001 or NIST SP 800-53 are inadequate and do not reflect the uniqueness of the nature of ICS/SCADA system. However, the safety requirements of IEC 61511 are matching practically with common security controls, and safety should be included in any new ICS/SCADA system security program. Henceforth, the researchers have concluded that a new security objective based on confidentiality, integrity, availability, and safety is required in ICS/SCADA system (Park and Lee 2014).

5.8 ICS/SCADA Security Requirements

5.8.1 Importance of Security Countermeasures

Security requirements (countermeasures) are an action, device, procedures, or techniques that reduce a threat, vulnerability, or an attack. These requirements are aimed to eliminate or prevent the threat by decreasing the negative impacts automatically or by discovering and reporting it so that corrective solutions can be taken by the IT specialists or the operators (D'Arcy et al. 2008). Department of Homeland Security of USA has developed several ICS/SCADA security countermeasures with related activities that are required to implement them, such as security policy, physical and environmental security, configuration management, communication protection, security awareness and training, incident response, portable media protection, access control, and security program governance (DHS-USA 2011).

Hentea (2008) has recommended several essential security requirements which must be implemented to secure the ICS/SCADA systems, including developing robust security policies and procedures, emerging a security knowledge management practices, so all ICS/SCADA stakeholders will be aware of how it is important to keep the whole system secure, and this security knowledge can be shared through formal training, awareness, meeting, etc. Furthermore, it is must to ensure that the

ICS/SCADA developer is highly qualified and skilled to create these critical systems and how to maintain the optimum security codes, and the various instruments which are installed and interconnected within the ICS/SCADA infrastructure should have inbuilt security feature for further security improvement and to create extra protection layer.

Protecting the field instruments (e.g., sensors, actuators, PLC, and RTU) is the main security concern for securing the operations of the ICS/SCADA system. This protection starts from selecting the right device with compliance to the latest security standards. Also, security of field instrument should go throughout its life cycle: procurement, installation, monitoring, and maintenance. Additionally, field instruments should have met the core security requirements such as the authentication, authorization, and CIA (Hentea 2008). Physical security countermeasures for field instruments should start from providing the applicable secure fences and gates which must be accessed through the combination of traditional hardware and electronic locks. Also, a motion detector linked with closed circuit television (CCTV) is recommended to record and discover any intruder to the remote stations.

Smart grid's field instruments and ICS/SCADA system must be protected from cyber-threats by adopting the right and effective countermeasures (Jokar et al. 2016). Protection must be utilizing the best security practices, such as preparing a formal and comprehensive ICS/SCADA security roadmap and the security policy should be well planned and adequate including the physical and logical access control. In addition, the design of the ICS/SCADA security and network architecture must be adequate. Furthermore, the security perimeter should be well defined, implemented, followed, and regularly updated. Also, other requirements are must including regular ICS/SCADA security audits and assessments, well-designed security configuration, OS and application security, and an adequate password policy which must be strictly implemented and followed. Security monitoring techniques must be having high-quality standards and functions. ICS/SCADA users must be strongly authenticated, and the data traveling between ICS/SCADA components should be well protected (Chae et al. 2015).

ICS/SCADA control room's servers and hosts must be well protected from the logical and physical attacks. Moreover, to provide the optimum availability the redundancy and high-capacity links between the servers must be established. It also recommended establishing load balancing and failover mechanisms between ICS/SCADA servers and network components, such as switches/routers (Kucuk et al. 2016). ICS/SCADA servers should receive the appropriate configuration/patch management and secure importation and execution of only trusted patches which will help to secure the ICS/SCADA servers. Furthermore, these servers require a strong multifactor authentication and limit the privileges to only employees to whom they need to access (Shahzad et al. 2014). Application whitelisting is considered as a good technique which must be implemented for ICS/SCADA servers and hosts to prevent installation and execution of unwelcomed applications (Larkin et al. 2014). To conclude, the ICS/SCADA servers and hosts should be secured physically, and logically through adopting and implementing solid access controls using different tools, policies, and protocols to develop the most secure identification, authentication, authorization, and accountability mechanisms (Fovino 2014).

In addition to the above, the following subsections will highlight some other important security key requirements that are needed to protect the ICS/SCADA systems.

5.8.2 Electronic and Physical Security Perimeters

To help to safeguard the physical and human assets of the ICS/SCADA infrastructure, there should be well-planned electronic and physical security perimeters. These security perimeters must be able to identify and control the people's access that enters and exit the ICS/SCADA facility, specifically the restricted areas. The electronic perimeters must be able to efficiently track the most current location, movements, and activities of the ICS/SCADA building occupant and assets, particularly in the event of any incident that might occur and must be able to respond quickly and raise alarms in real time whenever it is required (Fielding 2015).

One of the important electronic perimeters is the CCTV to monitor any physical entering the ICS/SCADA building specifically the control room. In addition, the type and number of CCTVs are required to be designed and determined by skilled professionals and security vendors as they are experts at the physical security countermeasures requirements for the ICS/SCADA infrastructure. Furthermore, purchasing an expensive or cheap CCTV is not always the practical and efficient, and balancing the scales of security against money is a tough process that leads to a confused decision. However, in terms of securing ICS/SCADA components buildings, it is important to conduct deep research of what type, specification, capabilities, and the best installation locations. For instance, providing CCTV with the type of electronically adjustable zooming/pan–tilt–zoom (PTZ) is a good feature to enhance the monitoring of the physical movements within the organization. Moreover, the combination of PTZ, security guards active monitoring, and motion-sense detectors is recommended in hardening the security of ICS/SCADA firm various buildings and departments (Knapp and Langill 2014).

Proper physical barriers and electronic access control devices must be installed to restrict access to the ICS/SCADA components facilities. However, high-tech incorporated solution is more recommended to secure the ICS/SCADA components facilities well. This unified security package should consist of video surveillance, access control, and perimeter detection tools that are tightly integrated with alarms and events raised in process and safety systems for intelligent, and a coordinated responses mechanism. The unified tool also should occupy with the human–machine interface (HMI) which give common look and feel results. The HMI is enhancing the decision-making because both ICS/SCADA control room consoles and security office desktops have visibility of the real-time events and lead to faster responds and feedbacks (Fielding 2015).

ICS/SCADA security professionals recommended adopting "Mantrap" technique to secure the control room. Mantrap is a small room meant to "trap" those who are willing to pass in this secured room. The mantrap is useful for giving slot time to verify the credentials of the individuals and either permit to access or raise alerts indicating an unauthorized entry. The suggested mantrap room should have two doors; the first is externally linked to the corporate common area, and the second is away from mantrap toward the corridor which leads to ICS/SCADA system control room and the authentication that is required for both doors. This room should occupy only one individual at a time for authentication purpose, and, if happened, there are more than one which raises notice requesting to leave the second person in the mantrap room. The detection of how many persons inside the mantrap room can be accomplished using motion detection technology (Niles 2004).

Most people characteristics tend to forget or neglect which might lead to not following the required security countermeasures; therefore, keeping warning employees and visitors by sticking warning posters in the critical places is good security practice. Normally, the statement of warning will be such as "only authorized use is allowed" and "device is being monitored". These banners should not disclose any technical and operational details about the device or system. There are many warning poster types targeted different situations which are used widely to warn employees, vendors, and visitors to understand risks and how they should be dealt with. Also, the most important aspect places the posters in the places that are noticeable by individuals which keep reminding them continuously, so no chance to neglect or forget them (Green 2001).

5.8.3   Network Communication Security

ICS/SCADA servers and network components work together to manage all communications, evaluate and examine the received data, and show the alerts and events on the HMI workstations (Chandia et al. 2008). Also, smart grid ICS/SCADA networks are depending on a hybrid of communication technologies including wired such as fiber optics, power line communication, copper-wire line, and wireless such as GSM/GPRS/WiMax/WLAN and Cognitive Radio (Yan et al. 2013). However, one of these communication technologies could represent a potential threat to the ICS/SCADA system (Babu et al. 2017). For instance, using old remote connection technology such as leased-line and dial-up modems for ICS/SCADA remote communication with distant field devices will facilitate the attacking efforts into the critical networks (Byres et al. 2007). These old communication technologies mostly have got no or weak authentication and encryption which can be utilized by attackers to gain access to the ICS/SCADA network. For example, an attacker can access to the old technology modem which is connected to the smart grid's breakers, disrupting the altering the control configuration settings causing power outages and damages various electrical equipments (NIST 2011).

From the above, these communication technologies must be well protected and secured, and developing a distinct policy is one of the important aspects to protecting the ICS/SCADA network. This policy describes traffic map or whitelisting which is essential for better governing the ICS/SCADA network by defining the data types and all allowed routes. The traffic whitelisting policy will be used to develop and maintain packets filtering and describe the potential bottlenecks locations (Barbosa et al. 2013). Also, the network policy should control the access to the network to protect the ICS/SCADA network infrastructure, by providing the required guidance to the ICS/SCADA users to their rights and what they can access and do, for instance, authenticating is required to make sure that the right person can access the right network's domain (Valenzano 2014).

In addition, most of the wireless network technologies remain with the default installation security setting and no hardening implemented to them and, consequently, they will be vulnerable to various attacks. Therefore, network policy should cover also the security of the wireless communications which should plan carefully bear in mind the noisy electrical environment of smart grid's substations which might impact the availability and reliability of the ICS/SCADA network (Leszczyna et al. 2011).

To conclude, it is important to utilize network security monitoring (NSM) and security information and event management (SIEM). NSM is a mean of gathering, analyzing, and rising the right indications or warnings to detect and respond to intrusions attacks (Sanders and Smith 2013). SIEM system is similar to NSM in providing a real-time analysis and performing data aggregation, association, alerting, dashboards, compliance, maintenance, and forensics investigation (Bhatt et al. 2014).

5.8.4 Product, Software, and Hardening

The ICS/SCADA infrastructure occupies a wide range of IT products and software, which should be evaluated and certificated to ensure the quality and to ensure that they are per the standing security standards. Also, these IT-related products and software should be updated, stable versions, tested, verified, and patched regularly to solid the security of the OT operational requirements. Similarly, security professionals should be aware of what functionalities and services are opened by default, and which should be disabled because most of them are not required, and the other should be more secured. While integrating new product or software in ICS/SCADA infrastructure, the security professionals must play a part in ensuring the integrity of the installed software, the verification of the software source (vendor, contractors), and the right and secured installation practice and environment (Krotofil and Gollmann 2013).

In addition, great care should be taken while add-ons, upgrading, updating, or patching any software work in the operational ICS/SCADA network. Also, these activities should be done only from a central server protected by a strong cryptographic technique while transiting. Furthermore, ICS/SCADA network product and software must be frequently updated and patched; and integrity should be preserved throughout the various upgrades. Also, the upgraded product and software should be offline to prevent causing any disruption to the ICS/SCADA network operational activities. Backup and recovery plan for ICS/SCADA products and software should be set for any bad events in advance. The plan should guide the security employees of performing routine latest backups of the running and startup configuration, data, software images, and modules. Also, the plan should record updated and tested recovery procedures, with consideration of no negative effect happened to ICS/SCADA network while restarting one of its components for system recovery purpose (Ramachandruni and Poornachandran 2015).

Hardening of ICS/SCADA system involves eliminating idle, needless, or unknown components such as modules, services, or ports. The most important in system hardening is selecting and implementing the most secure configuration parameters, as well as installing the security patches (Leszczyna et al. 2011). In addition, because the system hardening is very important, it should be taken with a great technical care. If IT specialist has got limited knowledge, it will produce a weak system hardening practices, and consequently, the overall ICS/SCADA system will have a weakness that might cause instability. For instance, an ignorant security specialist might remove a service that is rarely used, but it is important to serve other critical function. Thus, well training and technical awareness should be frequently conducted, so the concern employees will know how to perform good, stable, and strong ICS/SCADA system hardening. Also, this should be carried out, documented, and reviewed regularly with manufacturers, vendors, and contractors' collaboration. Moreover, this documentation should specify how and when ICS/SCADA system hardening is to be implemented or has been implemented (Graham et al. 2016).

Also, a good and systematic configuration and patch management strategy will facilitate hardening security plan and access control policy (Valenzano 2014). Moreover, the security core hardening configurations should not be changed during patching deployment process; and the most important is that asset's configuration should be verified during its lifecycle, which will help knowing when and which patch is required (Knowles et al. 2015).

Moreover, firewalls have a great role in minimizing the security risks for the ICS/SCADA system like the preventing from connection to the internet. Therefore, it is essential to harden the firewall to control the connection between ICS/SCADA and Internet aiming to stop unwanted inbound traffic. Moreover, solid hardening of the firewalls will restrict or control access from the corporate network to the ICS/SCADA network using well-defined access control lists (ACLs) and virtual local area networks (VLANs) access control policy (Stouffer et al. 2013).

5.8.5 *Portable Hardware/Devices Security*

Portable media such as USB devices, CDs, and DVDs will facilitate the intrusion by allowing attacker/hacker to access to ICS/SCADA system from inside the smart grid environment. For instance, USB stick might contain harmful malware or even be a booby-trapped that causes disaster to the ICS/SCADA system. In addition, portable devices are including (but are not restricted to) laptops, tablets (e.g., iPad), smart/mobile phones, digital camera/recording devices, personal digital assistants (PDAs), wireless keyboards and mice, and smartwatches. All these devices have ICT capabilities so they can be used to capture, store, processes, transmit, manage, and control, the data/applications which might create security threats to the ICS/SCADA system. Hence, ICS/SCADA system must be well secured and protected from portable devices by planning, developing, implementing, following, and continuing updating the security policy for these portable devices. For example, the policy should not allow any portable to be connected to the ICS/SCADA network unless it is strongly needed. Also, if it is required to do so, the portable device must be the first scan with the updated and well-known antivirus and malware application before connecting it to the secure network. A secure file transfer network zone is a good practice, so the files can be only exchanged through this network zone; hence, attaching portable device into the main ICS/SCADA network is prohibited and prevented risks which coming from booby-trapped USB devices for example (Alcaraz et al. 2012).

5.9 ICS/SCADA Security Policies

The objective of ICS/SCADA security policy is to provide management direction and support security professionals in an alignment of the business requirements and relevant laws and regulations. Because of the uniqueness of the critical ICS/SCADA environment, the security policy must be supported and approved by the decision-makers and distributed among the concern employees and stakeholders. The security policy document should include the individuals who are responsible for the implementing and managing the policy terms, and they should be identified by their names, position details, and the contacts information. Furthermore, the security policy should be systematically reviewed and updated to ensure its continuing suitability, adequacy, and effectiveness (ICTQatar 2012).

It is important to consider the well-known standards, guidelines, and best practices as guidance when needed to develop an effective ICS/SCADA security policy (Evans 2016). This policy should be annually reviewed and should conduct changes whenever is required bearing in mind that it will remain effective, suitable, and adequate (ICTQatar 2012).

ISO27001 Standard recommended several security policies such as Clear Desk and Clear Screen Policy, Access Control Policy, Disposal of Information/Media/ Equipment Policy, Data Classification and Control Policy, Mobile Computing and Teleworking Policy, Password policy, Penetration Testing Policy, System/Data Backup and Recovery Policy, Physical Security Policy, System Usage Monitoring Policy, Third Party Access Policy, and Virus /Malware Policy (Evans 2016).

5.10   Governance and Compliance

5.10.1   Governance of the ICS/SCADA Environment

Implementing the ICS/SCADA systems without proper planning processes with a poor risk management study, weak security countermeasures, and inappropriate access control, then it will lead to ICS/SCADA security program with unsuitable governance. In addition, one of the ICS/SCADA challenges is governance of OT (ICS/SCADA) security and IT security that are typically managed by different professionals. The OT devices are managed by the controlling, operating, engineering, or automation department, at the same time as the IT components are maintained by the IT department. Moreover, without good coordination, there is often a doubt about which department is responsible for the security of ICS/SCADA infrastructure which might lead to serious gaps in the organization's security competencies. Figure 5.2 illustrates the tasks, functions, and responsibilities of each department (IT and OT) for better managing the corporate and ICS/SCADA networks.

It is very important to align the OT and IT security strategies with overall business objectives and goals. The proper security alignment will enhance the resources allocation in a competent and operative manner to achieve the reliable business outcomes which must be measurable, and comparable with the risk assessment report. In fact, it will be negative consequences for the critical operations if the organization failed to develop good ICS/SCADA security governance. In addition, integrating well-known security standards, best practices, frameworks, and guidelines will give good guidance to develop a satisfactory governance structure for the security of ICS/SCADA infrastructure. In addition, the developed governance policy will help the decision-makers to make a better view of the ICS/ SCADA possible threats. It leads to how to mitigate them, enhance the stakeholder's internal communications and resources optimization, and clear the roles and responsibilities for the OT and IT professionals (Alcaraz and Zeadally 2015).

Official governance for the management of ICS/SCADA security will help to ensure that stakeholders and departments follow a steady and appropriate security strategy. Besides, the governance delivers clear roles and responsibilities, latest and up-to-date approaches to manage ICS/SCADA security threats, and guarantee that the supportive standards, guidelines, and policies are appropriate and are being

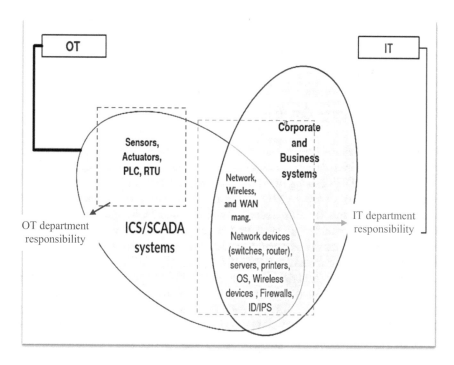

Fig. 5.2 Responsibilities of IT and OT to maintain and manage the corporate and ICS/SCADA networks

implemented and followed. In addition, governance documents should continuously review and make sure that ICS/SCADA security improvement is continuously adopted to reflect the changes in the organization infrastructure, objectives, and goals (Group 2015).

There are many activities that will help to governance the security of the ICS/SCADA such as business impact analyses for ICS/SCADA security, ICS/SCADA security classification, security roles, tasks and responsibilities, security awareness and training, managing the human factor, change management, exception handling, penetration test of the ICS/SCADA, incident response, business continuity, disaster recovery plans, ICS/SCADA asset management, and ICS/SCADA security report (CabinetOffice-UK 2014).

Adopting a good information security management system (ISMS) for ICS/SCADA systems specifically in smart grid is essential for a business to grow, innovate, and expand the knowledge base having highly confidential information. ISMS is used to handle information security risks by means of a group of policies, processes, procedures, and plans. In addition, ISMS is utilized to recognize threats and assign the applicable controls to manage or decrease them. ISMS has the required flexibility to adapt the security controls to any area of the organization and to improve the trust of the employees and stakeholders and ensure then that the valuable data is protected. In addition, the ISMS comprehensive IT-related risks security plan should cover both the corporate and ICS/SCADA systems (Knowles et al. 2015).

5.10.2 Regulatory Compliance and Standard Requirements

Several standardization bodies and governmental agencies have developed various standards, guidelines, and policies to protect ICS/SCADA systems. One of these guidelines is the DHS catalog of control systems security, which was developed by Homeland Security of USA (2011). This guideline is meant to guide the organizations to select and develop security controls, standards, guidelines, and best practices. Also, it provides guidance to determine the efficiency of the current ICS/SCADA security controls and whether they are compliance with the established security policy and procedures (DHS-USA 2011).

Furthermore, North American Electric Reliability Corporation (NERC) Critical Infrastructure Protection (CIP) Standards 002-009 and National Institute of Standards and Technology (NIST) Special Publication 800-82 have good guidance to ICS/SCADA security (NIST 2011). Also, NIST organization has developed the "NISTIR 7628 Guidelines for Smart Grid Cyber Security".

Table 5.1 shows some examples of the well-known ICS/SCADA security guidance at various levels of documents such as standards, guidelines, frameworks, policies, and regulations. For instance, when required to build good access control procedures, the security specialists can utilize the standards: BP1-CPNI, NIST SP 800-63, NIST SP 800-82, NIST SP 800-92, ISO/IEC 27001, ISO 15408, and ISO 19791. To build security requirements for the asset classification and control, we can utilize ISO/IEC 27001, ISO/IEC 27002, ISO 15408, and MAGERIT. In addition, top management when require building a plan to manage the business continuity they can use NIST SP800-34, NIST SP800-100, ISO/IEC 27001, ISO 15408, and MAGERIT standards.

Developing a security policy require intensive efforts and skills guided by standards such as NIST SP800-100, ISO/IEC 27001, and ISO 15408. Segmenting network is good security practice which can be created by following standards highlighted in Table 5.1 including BP1-CPNI, IEC 62351, and NIST SP 800-82.

Also, Table 5.2 summarized some of NERC CIP standards where each describes a security topic that smart grid needs to be addressed. For example, the standards (NERC CIP 001) talking about the sabotage reporting which aims to give guidance of the best way reporting disturbances or unusual occurrences to concerned authorities. Also, (NERC CIP 002) describes the proper method for identification and documentation of critical cyber assets using risk-based assessment methodologies. The standard NERC CIP 006 gives guidance to the security professionals of creation and maintenance of physical security controls, including processes, tools, and procedures to monitor perimeter access.

Table 5.1 Example of various standards describing different ICS/SCADA security topics

	BP1-CPNI	IEC 62351	NIST SP 800-34	NIST SP 800-63	NIST SP 800-76	NIST SP 800-82	NIST SP 800-92	NIST SP 800-48 & SP 800-97	NIST SP 800-100	ISO/IEC 27001	ISO/IEC 27002	ISO/IEC 27006	ISO 19011	ISO 15408	ISO 19791	MAGERIT
Access control	✓			✓		✓	✓			✓	✓			✓	✓	
Asset classification and control			✓							✓	✓			✓	✓	✓
Business continuity management									✓	✓	✓			✓	✓	
Certification procedures and audit									✓	✓		✓	✓	✓	✓	✓
Dial-up modem access	✓															
Differences between traditional IT and OT (ICS/SCADA)						✓										
Field technicians	✓															
ICS/SCADA characteristics, threats and vulnerabilities						✓										
ICS/SCADA security controls (management, operational, technical)						✓										

(continued)

Table 5.1 (continued)

	BP1-CPNI	IEC 62351	NIST SP 800-34	NIST SP 800-63	NIST SP 800-76	NIST SP 800-82	NIST SP 800-92	NIST SP 800-48 & SP 800-97	NIST SP 800-100	ISO/IEC 27001	ISO/IEC 27002	ISO/IEC 27006	ISO 19011	ISO 15408	ISO 19791	MAGERIT
Multi-connections to ICS/SCADA network	✓															
Network architecture security	✓															
Patch management strategies	✓					✓										
Physical and environment security										✓	✓			✓	✓	
Physical and logical DMZ	✓					✓										
Remote access password policy	✓															
Security policy									✓	✓	✓			✓		
Segmentation (VLAN or physical)	✓	✓				✓										
Wireless network security								✓								

Table 5.2 NERC CIP's standards each describes a security topic that smart grid needs to be addressed

	NERC CIP 001	NERC CIP 002	NERC CIP 003	NERC CIP 004	NERC CIP 005	NERC CIP 006	NERC CIP 007	NERC CIP 008	NERC CIP 009	NERC CIP 010	NERC CIP 011
Sabotage reporting	✓										
Critical cyber asset identification		✓									
Security management controls			✓								
Personnel and training				✓							
Electronic security protection					✓						
Physical security program						✓					
Systems security management							✓				
Incident reporting and response								✓			
Recovery plans for critical cyber assets									✓		
Bulk electrical system cyber system categorization										✓	
Bulk electrical system cyber system protection											✓

5.10.3   The Planning Phase to Protect ICS/SCADA

Preparing a comprehensive security program to secure the ICS/SCADA system is not an easy task because it involves so many procedures and steps. IT security specialists must coordinate with all stakeholders to prepare for this plan as it is a difficult task and consumes a lot of time and efforts, specifically if it is starting from scratch. This plan should be well prepared, organized, and managed to produce a stable, robust, practical, understandable, systematic, and obtaining the most updated knowledge in ICS/SCADA security field (Stouffer et al. 2015).

Eric Byres (Byres and Cusimano 2012) suggests seven phases to develop a new security plan for the ICS/SCADA system, which are assessing the existing systems, documenting the policies and procedures, training the employees and contractors, segmenting the ICS/SCADA system network and security, controlling the access to the ICS/SCADA system, hardening the components of the ICS/SCADA system, and monitoring and maintaining the security of ICS/SCADA system. In addition, these phases should be a continuous effort and the outcome should continually be evolved. For instance, once it reached the final stage, it might raise/discover a problem or might add a new device, system, or features that need to be further understood and assessed. Therefore, the cycle should commence with same phases and so on (Byres and Cusimano 2012).

5.11   ICS/SCADA Systems Security in Smart Grid

Today's traditional electric grids are being modernized with the addition of ICT technology capabilities, such as more automation, communication, and advanced IT systems (Yilin et al. 2012). These modernized electrical stations are one of the implementations of the CPSs, which depends mainly on sensors and digital systems to allow the power and communication data travel in two-way flows rather in one way with the traditional grid. This two-way communication allows providing electricity and information in real time or near time, which makes the power delivery more automated and creates more advanced power distribution and delivery networks (Fang et al. 2012). On the other hand, these modernized capabilities of the smart grid have added more security worries, such as DoS, human, and environmental attacks, breaking the confidentiality and integrity of data, or even making the power service totally unavailable (Florian et al. 2012). One of the security worries that most of smart grid ICT and embedded devices are located in different distance locations outside the utility premises, which make it difficult to be protected (Tan et al. 2017).

Smart grid should achieve the three triangle security objectives which include AIC. Availability is meant to ensure timely and reliably accessing to various critical smart grid operations and services, and failing to achieve that will cause severe disruption that might lead to different problems such as losing the power delivery.

Integrity objective is to keep SG data unmodified or destructed to ensure information non-repudiation and legitimacy. Confidentiality is aimed to design, develop, and manage a compact authorization mechanism for accessing the SG functions, services, and information to prevent illegal exposing of information that is not intended to be seen by public and individuals (Wanga and Lua 2013).

Historically, there were many attacks against the smart grid such as what happened in 2015 with the Ukrainian's regional electricity distribution company, which reported power service outages due to intruder into the ICS/SCADA systems which kept approximately 225,000 customers without power across various Ukrainian's locations for few hours (Sullivan and Kamensky 2017).

The new capabilities of smart grid have introduced security issues specifically in communication between field instruments back to ICS/SCADA control room, which opened the world's eye to the need to provide a secure encrypted and authenticated data exchange between these remote instruments and the ICS/SCADA host platform. Also, ICS/SCADA is considered to be a critical mission system in smart grid, and any disruption due to direct or indirect attack could be resulted in a considerable power disaster either financially, losing important data, physical destruction, or even it might lead to the loss of human life (Coates et al. 2010). Several studies have revealed that security of ICS/SCADA systems still weak and have many vulnerabilities, and failing to protect it might result in serious consequences for citizens and society (Luiijf et al. 2011).

Also, these modern technologies of the smart grid such as phasor measurement units (PMU), wide area measurement systems, substation automation, and advanced metering infrastructures (AMI) rely mainly on cyber resources which have great chance to be vulnerable to attack. Moreover, many past attacks incidents revealed that attackers use very intelligent methods against ICS/SCADA systems, and many countries have admitted that cyberattacks have targeted their critical infrastructure specifically the smart grids. Hence, various standards have been developed for technology, practices, and procedures which are used as a guidance of how to secure the SCADA system. Also, these practices and procedures take account of user training, host access, and how to react when ICS/SCADA security has been breached (Schneider 2012).

Information security investigations conducted by U.S. Government Accountability Office (GAO) revealed the weakness of the current ICS/SCADA security posture in the USA and NERC has recognized the result of these investigations and announced compliance requirements to enforce baseline cybersecurity efforts all over the bulk power system specifically the smart grids (Sridhar et al. 2012).

Environmental security issues can affect the performance of the ICS/SCADA systems. For instance, the ICS/SCADA control room components, data center devices, and field stations' instruments should be placed in a filtered environment to avoid the dust which might be conductive or magnetic that affects internal electronic parts. Also, temperature and humidity must be controlled and continuously monitored to protect the various electronic instruments specifically from overheating condition. In addition, visual and sound alarm technology should be adopted to trigger any negative environmental condition within ICS/SCADA infrastructure. Moreover, heating, ventilation, and air conditioning (HVAC) systems for ICS/SCADA control rooms must be well designed, implemented, monitored, and maintained. Additionally, fire controlling systems must be wisely planned to avoid initiating false or harmful consequences. Providing consistent power for ICS/SCADA control room, data center, and field stations' instruments is a crucial aspect. In addition, installing an emergency generator or/and an uninterruptible power supply (UPS) is also required in case of the main utility power failure. To conclude, modern smart fire prevention and HVAC systems must be protected against cyberattacks (Stouffer et al. 2015).

5.12   Conclusion

This chapter discusses the importance to keep ICS/SCADA systems protected as much as possible. Also, CPSs have many applications including the critical infrastructure organization such as the smart grid. In addition, we discussed how it is vital to keep smart grid secured and failing to do so will have great negative significant. The smart grid's operations are monitored and controlled using an ICS/ SCADA system which also must be protected and secured. The security of ICS/ SCADA system must be achieved through providing optimum objectives including AIC. This chapter also discussed the importance of adopting the right and proper security countermeasures to achieve the ICS/SCADA system objectives. Furthermore, this chapter went through several security requirements and how these will be applied in integrated IT and OT environment. Finally, it described various steps that will help to design, implement, and maintain security program dedicated for ICS/SCADA system working in the smart grid environment.

References

Aditya, S. G. P. M. (2015). *Aligning cyber-physical system safety and security*. iTrust—Center for Re-search in Cyber Security, Singapore University of Technology and Design.

Alcaraz, C., Fernandez, G., & Carvajal, F. (2012). Security aspects of SCADA and DCS environments. *Critical Infrastructure Protection*, 120–149.

Alcaraz, C., & Zeadally, S. (2015). Critical infrastructure protection: Requirements and challenges for the 21st century. *International Journal of Critical Infrastructure Protection, 8*, 53–66.

Ali, S., Anwar, R. W., & Hussain, O. K. (2015). Cyber security for cyber-physical systems: A trust based approach. *Journal of Theoratical and Applied Information Technology, 71*, 144–152.

Alshami, E., Albustani, H., & Melhem, M. (2008). Using supervisory control and data acquisition (SCADA) system in the management of a diesel generator. *Tishreen University Journal for Research and Scientific Studies—Engineering Sciences Series, 30*, 1–27.

Amin, S., Schwartz, G., & Hussain, A. (2013). In quest of bench-marking security risks to cyber-physical systems network. *IEEE, 27*, 19–24.

Babu, B., Ijyas, T., Muneer, P., & Varghese, J. (2017). Security issues in SCADA based industrial control systems. In *2nd International Conference on Anti-Cyber Crimes (ICACC)* (pp. 47–51). IEEE.

Barbosa, R. R. R., Sadre, R., & Pras, A. (2013). Flow whitelisting in SCADA networks. *International Journal of Critical Infrastructure Protection, 6,* 150–158.

Bhatt, S., Manadhata, P. K., & Zomlot, L. (2014). The operational role of security information and event management systems. *IEEE Security and Privacy, 12,* 35–41.

Byres, E., & Cusimano, J. (2012). *7 Steps to ICS and SCADA security.* (1st ed.). Tofino Security, exida Consulting LLC.

Byres, E., Leversage, D., & Kube, N. (2007). Security incidents and trends in SCADA and process industries. *The Industrial Ethernet Book.*

Cárdenas, A. A., Amin, S., Sinopoli, B., Giani, A., Perrig, A., & Sastry, S. (2009). Challenges for securing cyber physical systems.

Cabinetoffice-UK. (2014). Government security classifications. UK: www.gov.uk.

Chae, H., Shahzad, A., Irfan, M., Lee, H., & Lee, M. (2015). Industrial control systems vulnerabilities and security issues and future enhancement. *Advanced Science and Technology Letters, 95*(CIA 2015), 144–148.

Chandia, R., Gonzalez, J., Kilpatrick, T., Papa, M., & Shenoi, S. (2008). Security strategies for SCADA networks. In: E. Goetz & S Shenoi (Eds.), *Critical infrastructure protection.* Boston, MA: Springer US.

Charalambos, K., Michail, M., Fareena, S., Shiyan, H., Jim, P., & Yier, J. (2015). Cyber-physical systems: A security perspective. In *2015 20th IEEE European Test Symposium (ETS).* IEEE European Test Symposium (ETS).

Coates, G. M., Hopkinson, K. M., Graham, S. R., & Kurkowski, S. H. (2010). A trust system architecture for SCADA network security. *IEEE Transactions on Power Delivery, 25,* 11.

D'arcy, J., Hovav, A., & Galletta, D. (2008). User awareness of security countermeasures and its impact on information systems misuse: A deterrence approach. *Information Systems Research,* 1–20.

Darktrace. (2015). Cyber security for corporate and industrial control systems. *Darktrace Industrial Immune System Provides Continuous Threat Monitoring for Oil & Gas, Energy, Utilities, and Manufacturing Plants.* USA: Darktrace Limited. www.darktrace.com.

Dell. (2015). 2015 Dell security annual threat report. In D. Inc (Ed.) USA.

DHS-USA. (2011). Catalog of control systems security: Recommendations for standards developers. In *Control Systems Security Program.* USA: U.S. Department of Homeland Security.

Evans, L. (2016). Protecting information assets using ISO/IEC security standards. *Information Management, 50,* 28.

Fang, X., Misra, S., Xue, G., & Yang, D. (2012). Smart grid—The new and improved power grid: A survey. *IEEE Communications Surveys & Tutorials, 14,* 944–980.

Fielding, A. H. (2015). *Physical security for industrial assets* [Online]. International Society of Automation (ISA). Available: https://www.isa.org/intech/20151003/. Accessed 2016.

Florian, S., Zhendong, M., Thomas, B., & Helmut, G. (2012). A survey on threats and vulnerabilities in smart metering infrastructures. *International Journal of Smart Grid and Clean Energy (SGCE), 1,* 7.

Fovino, I. N. (2014). SCADA system cyber security. In *Secure smart embedded devices, platforms and applications.* Springer.

Galloway, B., & Hancke, G. P. (2013). Introduction to industrial control networks. *IEEE Communications Surveys and Tutorials, 15,* 860–880.

Gao, J., Liu, J., Rajan, B., Nori, R., Fu, B., Xiao, Y., Liang, W., & Chen, C. L. P. (2013). SCADA communication and security issues. *Security and Communication Networks* [Online], 7.

Graham, J., Hieb, J., & Naber, J. (2016). Improving cybersecurity for Industrial Control Systems. Paper presented at the 25th IEEE International Symposium on Industrial Electronics, Santa Clara, USA. 618–623

Green, M. (2001). The psychology of warnings. *Occupational Health and Safety Canada,* 30–38.

Group, P. A. C. (2015). Security for industrial control systems improve awareness and skills a good practice guide. Available: https://www.ncsc.gov.uk/content/files/protected_files/guidance_files/SICS%20-%20Improve%20Awareness%20and%20Skills%20Final%20v1.0.pdf.

Hentea, M. (2008). Improving security for SCADA control systems. *Interdisciplinary Journal of Information, Knowledge, and Management, 3,* 73–86.

Homeland-Security. (2009). Department of Homeland Security: Cyber security procurement language for control systems. In *Control system security program.* USA: Homeland Security.

ICTQatar. (2012). Controls for the security of critical industrial automation and control systems guidelines. In Q. N. I. Assurance (Ed.). Qatar: Qatar National Information Assurance.

Ida, N. (2014). *Sensors, actuators, and their interfaces: A multidisciplinary introduction.* SciTech Publishing Incorporated.

IEC, I. E. C. (2016). *IEC TC57 WG15: IEC 62351 security standards for the power system information infrastructure.* National Association of Regulatory Utility: Xanthus Consulting International.

Jokar, P., Arianpoo, N., & Leung, V. (2016). A survey on security issues in smart grids. *Security and Communication Networks, 9,* 262–273.

Khurana, H., Hadley, M., Lu, N., & Frincke, D. A. (2010). Smart-grid security issues. *IEEE Security & Privacy,* 81–85.

Kim, K.-D., & Kumar, P. R. (2013). An overview and some challenges in cyber-physical systems. *Journal of the Indian Institute of Science, 93,* 10.

Knapp, E. D., & Langill, J. T. (2014). *Industrial network security: Securing critical infrastructure networks for smart grid, SCADA, and other industrial control systems.* Syngress.

Knowles, W., Prince, D., Hutchison, D., Disso, J. F. P., & Jones, K. (2015). A survey of cyber security management in industrial control systems. *International Journal of Critical Infrastructure Protection, 9,* 52–80.

Krotofil, M., & Gollmann, D. (2013). Industrial control systems security: What is happening? In *11th IEEE International Conference on Industrial Informatics (INDIN), 2013* (pp. 670–675). IEEE.

Kucuk, S., Arslan, F., Bayrak, M., & Contreras, G. (2016). Load management of industrial facilities electrical system using intelligent supervision, control and monitoring systems. In *2016 International Symposium on Networks, Computers and Communications (ISNCC), 2016* (pp. 1–6). IEEE.

Kui, R., Zuyi, L., & Robert, C. Q. (2011). Guest editorial cyber, physical, and system security for smart grid. *IEEE Transactions On Smart Grid, 2.*

Kuzlu, M., Pipattanasomporn, M., & Rahman, S. (2014). Communication network requirements for major smart grid applications in HAN, NAN and WAN. *Computer Networks, 67,* 74–88.

Laing, C., BADII, A., & Vickers, P. (2012). *Securing critical infrastructures and critical control systems: Approaches for threat protection: Approaches for threat protection.* IGI Global.

Larkin, R. D., Lopez Jr, J., Butts, J. W., & Grimaila, M. R. (2014). Evaluation of security solutions in the SCADA environment. *ACM SIGMIS Database, 45,* 38–53.

Lee, R. M., Assante, M. J., & Conway, T. (2015). ICS CP/PE (cyber-to-physical or process effects) case study paper—German Steel Mill cyber attack. In SANS, I. C. S. (Ed.). SANS-ICS.

Leszczyna, R., Egozcue, E., Tarrafeta, L., Villar, V. F., & Alonso, J. (2011). Protecting industrial control systems. European Network and Information Security Agency (ENISA).

Luiijf, E., Ali, M., & Zielstra, A. (2011). Assessing and improving SCADA security in the Dutch drinking water sector. *International Journal of Critical Infrastructure Protection, 4,* 124–134.

Mo, Y., Kim, T.-H., Brancik, K., Dickinson, D., Lee, H., Perrig, A., et al. (2012). Cyber–physical security of a smart grid infrastructure. *Proceedings of the IEEE, 100,* 195–209.

Moscatelli, A. (2011). From smart metering to smart grids: PLC technology evolutions. In *15th IEEE International Symposium on Power Line Communications and its Applications.* University of Udine in Italy.

Niles, S. (2004). *Physical security in mission critical facilities.* Schneider Electric White Paper 82, Revision 2.

NIST. (2011). Guide to industrial control systems (ICS) security. *Recommendations of the National Institute of Standards and Technology*. USA: U.S. Department of Commerce.

NIST. (2015). Guide to industrial control systems (ICS) security. *NIST Special Publication 800-82 Revision 2*. USA: National Institute of Standards and Technology.

NIST. (2016). Framework for cyber-physical systems. National Institute of Standards and Technology.

Park, S., & Lee, K. (2014). Advanced approach to information security management system model for industrial control system. 348305. Available: https://www.ncbi.nlm.nih.gov/pmc/articles/PMC4129153/.

Rajkumar, R. R., Lee, I., Sha, L., & Stankovic, J. (2010). Cyber-physical systems: The next computing revolution. In *Proceedings of the 47th Design Automation Conference* (pp. 731–736). ACM.

Ralston, P. A. S., Graham, J. H., & Hieb, J. L. (2007). Cyber security risk assessment for SCADA and DCS networks. *ISA Transactions, 46*, 583–594.

Ramachandruni, R. S., & Poornachandran, P. (2015). Detecting the network attack vectors on SCADA systems. In *International Conference on Advances in Computing, Communications and Informatics (ICACCI)* (pp. 707–712). IEEE.

Sanders, C., & Smith, J. (2013). *Applied network security monitoring: Collection, detection, and analysis*. Elsevier.

Sanislav, T., & Miclea, L. (2012). Cyber-physical systems-concept, challenges and research areas. *Journal of Control Engineering and Applied Informatics, 14*, 28–33.

Schneider, E. (2012). SCADA systems. *Telemetry & Remote SCADA Solutions* [Online]. Available: https://www.schneider-electric.com.au/en/product-category/6000-telemetry-and-remote-scada-systems/. Accessed September, 2016.

Shahzad, A., Musa, S., Aborujilah, A., & Irfan, M. (2014). The SCADA review: System components, architecture, protocols and future security trends. *American Journal of Applied Sciences, 11*, 1418.

Shuaib, K., Trabelsi, Z., Abed-Hafez, M., Gaouda, A., & Alahmad, M. (2015). Resiliency of smart power meters to common security attacks. *Procedia Computer Science, 52*, 145–152.

Sridhar, S., Hahn, A., & Govindarasu, M. (2012). Cyber-physical system security for the electric power grid. *Proceedings of the IEEE, 100*, 14.

Stouffer, K., Falco, J., & Scarfone, K. (2013). Guide to industrial control systems (ICS) security. *National Institute of Standards and Technology (NIST)*. U.S. Department of Commerce, NIST Special Publication 800-82.

Stouffer, K., Pillitteri, V., Lightman, S., & Hahn, A. (2015). Guide to industrial control systems (ICS) security—Revision 2. In N. S. P. (Ed.), 800-82. United States of America.

Sullivan, J. E., & Kamensky, D. (2017). How cyber-attacks in Ukraine show the vulnerability of the US power grid. *The Electricity Journal, 30*, 30–35.

Tan, S., De, D., Song, W.-Z., Yang, J., & Das, S. K. (2017). Survey of security advances in smart grid: A data driven approach. *IEEE Communications Surveys & Tutorials, 19*, 397–422.

Thomas, M. S., & McDonald, J. D. (2015). *Power system SCADA and smart grids*. Boca Raton: FL, CRC Press.

Valenzano, A. (2014). Industrial cybersecurity: Improving security through access control policy models. *IEEE Industrial Electronics Magazine, 8*, 6–17.

Wanga, W., & Lua, Z. (2013). Cyber security in the smart grid: Survey and challenges. *Elsevier, 57*, 1344–1371.

Wu, F.-J., Kao, Y.-F., & Tseng, Y.-C. (2011). Review From wireless sensor networks towards cyber physical systems. *Pervasive and Mobile Computing, 7*, 397–413.

Yan, Y., Qian, Y., Sharif, H., & Tipper, D. (2013). A survey on smart grid communication infrastructures: Motivations, requirements and challenges. *IEEE Communications Surveys & Tutorials, 15*, 5–20.

Yilin, M., Tiffany, H.-J. K., Kenneth, B., Dona, D., Heejo, L., Adrian, P., et al. (2012). Cyber-physical security of a smart grid infrastructure. *IEEE, 100*, 15.

Chapter 6
Embedded Systems Security for Cyber-Physical Systems

Embedded systems have become a core component in today's technological era. Embedded systems are a combination of hardware and software specially designed to perform a specific task. They can control many common systems used today which are implemented in variety of field such as transportation, health care, communication, and many more. Embedded systems have also obtained features of functioning using the Internet and this is what is often referred to as cyber-physical systems (CPSs). This chapter aims to study the interrelationships between embedded systems and CPS. It also investigates the key security areas in both fields and highlights some challenges and gaps in these two fields with regards to their complex nature, trade-offs for efficient functioning of such systems and trust and reputation approach.

6.1 Introduction

Embedded systems can broadly be defined as a system that consists of a required computer hardware which runs using a coded software to perform an assigned task (Kamal 2008). Embedded systems consist of certain inputs, on which given operations are executed to obtain the desired output. Thus embedded systems resemble the characteristics of any traditional or classic systems. Embedded systems are one of the most vital components of the modern information technology (IT) age. The proof of the above statement is given by the extensive implementations of embedded systems in the fields of health care (Lee et al. 2006), transportation (Navet and Simonot-Lion 2008), intelligence (Alippi 2014), communication (Segovia et al. 2012), control (Zometa et al. 2012), governance (Shukla 2015), etc. It has led to modernization and electronization of fields that was difficult to achieve even a few decades back. Due to its endless advancements, it has become an integral part of human lives (Jalali 2009). The importance is further highlighted by the fact that, currently, over 95% of all the chips produced are for embedded

© Springer International Publishing AG 2018
S. Ali et al., *Cyber Security for Cyber Physical Systems*, Studies in Computational Intelligence 768, https://doi.org/10.1007/978-3-319-75880-0_6

systems (Sifakis 2011; Ma et al. 2007). The sale of 12 billion Advanced RISC Machines (ARM) chips in 2014 is yet another proof of the extensive use and production of embedded systems (PLC 2014). Some implementations of embedded systems are extended to products like home appliances (Nath and Datta 2014), automobiles, robots, airplanes, mobile phones, etc.

Physically, embedded systems are small computers assembled with a microprocessor combined with a small read-only memory (ROM) and some application-based essential peripherals. The embedded systems are specifically designed for a desired application. The functionality of the system is programmed using a code to perform the expected task. The concept of embedded systems was initiated with the development of the Apollo Guidance Computer in 1960 (Hall 1996). Within a year, embedded systems started its mass production for D17 Minuteman Missiles (Timmerman 2007) and within a decade came the first embedded computer, the Intel Company and the 4-bit microprocessor. Earlier, embedded systems were generally reactive systems, performing a single task assigned with no scope of updating or interacting to other systems. With the advancement in technology, embedded systems were able to interconnect with other systems to form distributed and parallel systems (Sharp 1986). These systems were capable of performing even more broad, effective, and efficient tasks. Later, implementing the extensive interconnectivity, the concept of everywhere and anytime was introduced in the 1990s by Weiser (1991) and Davies and Gellersen (2002). This introduction led to the ubiquity of the embedded systems and they were termed as ubiquitous systems. Since a decade, embedded systems have also obtained the functionality to work using the Internet and thus, the term "Cyber-Physical Systems" (CPSs) have come into existence.

The term CPS was coined in 2006 by Helen Gill of the National Science Foundation (NSF) in United States (Gunes et al. 2014; Lee and Seshia 2014). Since then, various advancements have taken place in this field which have reached new breakthroughs and heights. Just like its predecessor, the embedded systems, it brought about a broad range of applications in fields like health care (Lee and Sokolsky 2010), transportation (Wasicek et al. 2014; Osswald et al. 2014), power (Sridhar et al. 2012; Ashok et al. 2014), control (Backhaus et al. 2013; Parolini et al. 2012), communication (Fink et al. 2012), etc. It is easily noticeable that the CPSs are being implemented in various crucial fields and thus they form the critical infrastructures (Das et al. 2012; Miller 2014). Since CPS now forms a key part of the society, security of CPS becomes a prime concern for their smooth and safe functioning. The seriousness is proved by the security failures that have occurred in the past, leading to losses of finance, property, confidentiality, and/or lives. Attacks on a medical CPS may lead to fatal consequences and death (Talbot 2012; Mitchell and Chen 2015; Halperin et al. 2008), attacks on supervisory control and data acquisition (SCADA) system lead to malfunctions (Zhu et al. 2011a), e.g., Stuxnet (Farwell and Rohozinski 2011), attacks on transportation and aviation lead to loss of control of the machine and chaos (Denning 2000; Pike et al. 2015), e.g., Derailed tram in Poland (Leyden 2008), attacks on public industries like water distribution or pipeline system lead to fatal components into the system and accidents (Slay and

Miller 2008; Tsang 2010), e.g., Gas pipeline in Russia (Quinn-Judge 2002), attacks on critical infrastructures lead to total cessation to a city or country (Baker et al. 2009), e.g., Hackers cut city's electricity for extortion (Greenberg 2008), and for the latest, attacks on a smart grid lead to power cut and losses (Conti 2010). Realizing the severity of CPS field from the above mentioned issues, security forms the base of this book. This chapter discusses the concepts of embedded systems and security and describes its relationship with the CPS and the challenges involved.

6.2 Embedded Systems in CPS

The foundation of the CPS was laid by the embedded systems. In this section, the relationship between these two entities has been discussed. Marwedel, in his book (Marwedel 2010), clearly showed that the embedded systems are the backbone of the CPS. The book is completely dedicated to the designing of CPS using the embedded systems and discusses the requirements, challenges, constraints, applications, software coding, and all other concepts related to the CPS design. A modest description on security has also been provided, with discussions on the hardware and software requirements. Parvin et al, in their research on CPS, recognized that CPS is built on the research from disciplines like embedded systems and sensor networks (Parvin et al. 2013).

Lee in 2010 (Lee 2010) described CPS as an intersection of cyber and physical realms rather than just a union. Embedded systems were no longer to be used for specific purposes which are designed once and inflexible in its application. The modern embedded systems forming the CPS are larger systems with more electronics, networking, and physical processes. He discussed cyberizing the physical components and physicalizing the cyber parts of the systems.

In 2012, Broy et al. (2012) and Broy and Schmidt (2014) defined CPS as "the integration of embedded systems with global networks such as the Internet" and the first step toward CPS is identified as networked embedded systems. The authors listed out the other technologies like Internet as business, radio frequency identification (RFID), Semantic Web, and applications like Android, Firefox, etc., as the driving force toward CPS.

Magureanu et al. (2013) termed CPS as "massively distributed heterogeneous embedded systems linked through wired/wireless connections" and later as "embedded distributed systems".

Xia et al. (2011) realized the possible feasibility of the IEEE 80.15.4 protocol in CPS and thus conducted a performance evaluation for it using two different modes. They describe CPS as networked embedded systems that bridge the computing and communication of the virtual world to the real world. The major distinguishing factor between CPS and embedded systems is the fact that CPS can be a large system of wireless sensor and actuator network (WSAN) while the other can only be a system of WSN. The evaluation shows that the default configuration does not yield the best performance and thus CPS requires its own configuration and in the future, it may require new standards being defined.

Stojmenovic (2014), termed "control technologies with non-networked embedded systems are examples of CPS that are not IoT" and since CPS is also related to control systems, CPS can also be defined as the integration of computation, communication, and control processes. The current task of the researchers in this field is identified as building on embedded systems by integrating the software and networking with the physical environment.

Battram et al. (2015) focused on the contracts used during the CPS production and design. The noticeable difference between the two is that embedded systems are small computing devices deeply planted in the surroundings as such that their existence tends to be unknown while CPS takes the environment into account and thus emphasizes its much larger size as compared to the embedded systems.

A clear differentiation was provided by Sanislav and Miclea (2012). CPS is not the traditional embedded systems, real-time systems, the sensors networks, and desktop applications but a unique integration of them all. The prime characteristics of CPS are as follows:

- It has cyber capabilities;
- It is networked at a large scale;
- It is dynamic and reconfigurable;
- It includes higher extent of automation as compared to embedded systems;
- It executes many more computations and commands; and
- It has cyber and physical components which are more deeply integrated.

6.2.1 Transformation

One of the earliest papers discussing the transformation of embedded systems to CPS was by Lee in 2008 (Lee 2008). It described the design challenges toward CPSs. He claimed that in the past, embedded systems had always maintained a high reliability and predictability standard as compared to general purpose computing. The transformation to CPS will only lead to further increment in the expectation of reliability and predictability of the systems.

In 2011, Wan et al. (2011) and Shi et al. (2011) explained the relationship between the embedded systems and CPS. Though both terms are used interchangeably, they list seven advancements of CPS compared to embedded systems and WSN. They proved the importance of these systems by listing various funding and projects being carried out in US, Europe, China, Korea, etc. They also discussed issues like energy management, network security, data transmission managements, model design, control technique, and resource allocations.

Kim et al. (Kim and Kumar 2012), in 2012, highlight that the embedded systems are the fundamental components of the CPSs, as it is used in distributed sensing, computation, and control over various communication media, level, and algorithms.

The authors termed the CPS as the networked embedded systems. The paper also listed various projects being conducted relating to embedded systems and CPS, the security challenges in CPS and the applications.

Wan et al. (2013) studied the transformation of the technology from the machine-to-machine (M2M) communication to CPS. They present a broad view on this concept of the IT and group the CPS, WSN, and M2M under the umbrella of Internet of Things (IoT). CPS is considered as one of the highest levels of IoT and it has evolved due to the developments in the field of embedded systems. They discussed the applications like the cloud computing, unmanned vehicles, vehicular ad hoc network (VANET), mobile agent, distributed real-time control, etc.

Gupta et al. (2013) studied the potential in CPS and published their work describing the future of the technologies. Embedded system technology is identified as one of the major contributors to CPS technology since it not only provides the hardware but also the software, modeling, and designing tools. The authors also described the features and some future applications of the CPS.

Gurgen et al. (2013), in their publication on self-aware CPS for smart buildings, provided many recommendations toward self-manageable CPS. Among those, they point out that the fundamental research on CPS is from various fields including the embedded systems. Some of the features like robustness, reliability, dependability, effectiveness, and correctness are to be derived from the embedded system technology.

In the article by Bartocci et al. (2014), CPSs are termed as the next generation of embedded systems and the theoretical and practical challenges for CPS are discussed. Embedded systems which focused on design optimization in the past have shifted the focus to complex alliance and interaction between the computational devices and the physical environment and this lead to the formation of CPS.

Lee (2015c) discussed the past, present, and the future of CPS and Casale-Rossi et al. (2015) discuss the future of electronics and semiconductors in Europe. In both, the two terms are once again observed to be used interchangeably showing that with time, the basic embedded systems have been overshadowed and engulfed by the CPS and now both terms mean almost the same concept. On the contrary to this, Greenwood et al.(2015) presented a good work to differentiate between embedded systems and CPS in a clear way. They defined embedded system as "an information processing system where the end user is unaware of a computer being present" while CPSs are embedded systems that tightly link the computing resources to the physical world. The definition for CPS is further elaborated by Greenwood in the paper (Greenwood et al. 2015).

A good analysis and study on the transformation and maturation of embedded systems have been provided by Mosterman and Zander in one of the papers published (Mosterman and Zander 2015). The transformation is explained in three steps starting with networked embedded systems to a paradigm shift of the system and finally the formation of the CPS. The later part of the paper describes the challenges and the software involved in the design and function of CPS.

6.2.2 Modeling and Design

In 2008, during the initial stages of the CPS, Tan et al. (2008) proposed prototype architecture for CPS. The authors mentioned that the current technologies of embedded systems and the cyber-world integrate to form the CPS. The fundamental gaps between the embedded systems and CPS are explained and the prototype architecture for the CPS has been provided. Some of the requirements to model or design such systems are the global reference time to keep uniformity among the components. The systems must be event/information-driven and qualified to bring trust among the components and systems. These systems must publish/subscribe scheme based on the interest and goal of the CPS, control laws on the actions, and new networking techniques to make these possible to function smoothly.

Koubâa and Andersson (2009), in 2009, provided a connection between the Internet and the embedded systems. Internet is used to interconnect computers and share digital data while embedded systems are used to control systems in real time and a combination of the two formed the CPS. They discussed the concept of cyber-physical Internet by presenting its design requirements and limitations. The protocol stack architecture consists of six layers: physical, data link, network, transport, application, and cyber-physical layer. Some of the challenges faced were operating these new systems, designing new network protocols, real-time operations, effective performance, and data aggregation of all the information collected and generated.

Lee (2006, 2007, 2008) discussed the design challenges of the CPS. The embedded systems are identified as the basic building blocks of the CPS. To proceed into the CPS, rebuilding the computing and the networking abstractions was recommended. Tham and Luo (2013) presented a productive work on the feasibility of decision-making by prediction for purchasing electricity in smart grid environment. They researched on the smart grid CPS (SG-CPS) and examined two optimizations methods and noted that the performance of the predictions was improved in terms of execution time and memory requirements. For these kinds of CPSs, they implemented embedded systems based on ARM.

Karnouskos et al. (2014) proposed a cloud-based approach toward industrial CPS. The advanced features like multi-core processing, threading, memory on chip, etc., in CPS were identified as some of the major advancements from embedded systems and fusing cloud and CPS is expected to bring further advancement in this field. The term they used for this fusion is "Cloud of things".

To model and control the CPS, Bujorianu and MacKay (2014) proposed a framework based on complex science. Embedded systems are identified as one of the predecessors and the basics of CPS have been already studied through this paradigm. Concepts of modeling, control, and various abstractions have been discussed.

Once the designing and modeling of CPS is executed, the stage of validation and verification is implemented and a paper on this was presented by Zheng et al. (2014). To understand the state of the art, they used three methods, namely, questionnaire survey with researchers, interview with CPS developers, and literature review. One of the questions for the survey dealt with the differences between the CPS and embedded systems. CPS is found to be a defined by many as embedded systems that also include network, security, and privacy, cloud computing and big data, which are almost absent. Although researchers used the two terms interchangeably many time, the developers have a clear differentiation among the two.

Johnson et al. (2015) addressed the specification mismatch in CPS. They noted that the embedded systems are getting increasingly complex to evolve into the CPS. Embedded systems software is still being used to design and build the CPS. Various research and techniques have been identified that can address the issue of mismatch in CPS specifications by modifying these software to bridge the gap between the two concepts. A method and tool have also been developed for the same.

6.2.3 Tools and Languages

Examining the state of the art and the requirements, Lee (2006, 2007, 2008) pointed out challenges in the networking of embedded systems, adding an operating system (OS) into them and the reaction to multiple real-time inputs and outputs to sensors and actuators, respectively. Some of the tools mentioned are TinyOS and netC. He presented the four-layer abstraction of the CPS with hardware like FPGS, ASIC, and microprocessors and programming languages like Java, C++, and VHDL. The layers are task-level model, programs level, executable level, and the silicon chip level. He has further studied this field and presented a comprehensive work in the book (Lee and Seshia 2014). Yet another paper on the tools for CPS was presented by Zhou et al. (2014). The tools for hardware and software design have been discussed. The tools for embedded systems are determined as the foundation of designing any CPS. The tools like ALLEGRO, PROTEl, and PADS used for designing PCBs have been compared. Other hardware tools discussed are Proteus, Keil, ARM ADS, RealView MDK, Multi 2000, Quartus II, LabView, and Multisim. Some of the embedded OSs as mentioned above have also been discussed. The paper also gives an insight into the network analysis and simulation tools, since networking is a vital part of the CPS. Tools like Ns-2, OMNeT++, OPNET, SPW, J-sim, GloMosim, and SSFnet were listed and compared. Other tools discussed were codesign tools for real-time behavior and domain-specific tools.

Eidson et al. (2012) tried to address one of the challenges in the evolution of CPS and that was the designing of CPS. They discussed the use of the Ptolemy software, developed for modeling simulation and design of embedded systems at UC-Berkeley (Lee 2015a, b; Ptolemaeus 2014), for designing CPS. The PTIDES extension to the Ptolemy II software is discussed with its application to model CPSs. The model is an event-driven process and the applications discussed were that of the power plant control.

Magureanu et al. (2013), in 2013, discussed Unified Modeling Language (UML) used for modeling embedded systems, which is also being used for modeling the CPSs. The authors presented a good work on the modeling and verification of CPS using the UML, its various specifications like Z and PVS, and their differences. Finally, they discuss the validation of the static properties in the UML where two specifications mentioned.

Wan et al. (2013) argued that though there are tools like MATLAB and Truetime available for embedded system design, they are not satisfactory for the CPS design and thus required some improvements and innovations.

For designing of the CPS for energy management and control, Molina et al. (2014) conducted a project named SmartCoDe. Describing the complexity in CPS, the authors used a model-based approach to design the distributed embedded CPS using the System C tool.

Pohlmann et al., worked on the designing of CPS using the Modelica language (Pohlmann et al. 2014). They have signified that CPS is a collaborated form of many individual systems while embedded systems are systems that operate individually. They proposed and created new patterns for the collaboration and coordination activities in the CPS, which could be reused for various applications.

6.2.4 Applications

Koubâa and Andersson (2009) discuss the technologies involved like WSN, RFID, ZigBee, mobile phones, etc. which have merged into CPS.

Rajkumar et al. (2010) spotted the potential in the CPS and published their findings in 2010. They described CPS as the combination embedded systems, real-time systems, distributed sensor systems and controls. They have described some of the major applications of CPS in the advanced power grid, better disaster response and recovery, and various assistive devices.

Osswald et al. (2014) presented their prototype CPS for automotive. They recognized the extensive implementation of embedded systems in the automobiles since its introduction by Volkswagen in 1968. They stated that "the embedded systems are closed systems performing specific tasks and they are now being challenged by the transformation of these devices into an interconnected network". The authors proposed the human integration into the CPS cycle for better performance.

Keller et al. (2014) channeled their research on smart homes, which is again related to the energy management. They proposed the dual reality approach which enables the user to perform a similar action in both, physical and virtual worlds by synchronizing them. They explained that CPSs are network-compatible embedded systems that are created to form all-in-one solution to a particular problem. The dual reality approach solves the problems of ease of use, flexibility and interaction between the two worlds which leads to improved quality of life.

6.3 Challenges in Transformation

The addition of networking into the system converts an embedded system to a CPS, and it is the demand of the current technological age. The networking exposes the computing and control capabilities to the outside world, which has been confined within the system's interiors all these time.

Lee described the design challenges of CPS in 2008 (Lee 2008). The challenge of maintaining and improving the reliability and predictability observed in embedded systems has been discussed in previous sections. Another challenge was the rebuilding of the abstractions which has been discussed in Sect. 6.2. Maintaining the timing of commands and actions for the system to perform real-time application is yet another challenge. Bringing an upgrade into the critical infrastructures requires extensive testing and verification, without it may lead to catastrophes. In the same year, Bonakdarpour (2008) published a paper discussing the challenges in transforming from embedded systems to CPS. Some of the challenges were the scalability, abstraction, fault tolerance, hybridity, and implementation of formal methods.

One of the major challenges related to the embedded systems is the real-time abstraction of the information, as pointed out by Rajkumar et al. (2010). Since CPS is dependent on the distributed network, time synchronization among all the devices in the system is very vital for the feasible functioning of the system. They also mentioned that new architectural patterns, hierarchy, theories, protocols, languages, and tools must be devised to obtain the most efficient and effective CPSs (Rajkumar et al. 2010).

Broy et al. (2012) and Broy and Schmidt (2014) presented the imminent challenges of CPS in their publications. To demonstrate the evolution of embedded system into CPS, the brake system in vehicles is taken as an example. The challenges for CPS were divided into the technological and social aspects. Some of those challenges are human–system cooperation, integrated tools and systems, expansion of the quality and engineering standards, acceptance of the new technology, economic and research interest in CPS. Broy in 2013 (Broy 2013) describes the challenges from an engineering perspective. CPS is described as "the combination of embedded system with cyberspace". The major challenge in CPS is pointed out as combining the properties of embedded systems such as real-time, dependability, reliability, and functional safety with the openness and reduced availability and dependability of the Internet.

Raciti and Nadjm-Tehrani (2013) recognize the CPS as embedded system when considering the hardware limitations for higher security deployment as these systems must be kept cheap to fulfill the large-scale production criteria.

Fitzgerald et al. (2014), in 2014, presented a comprehensive work on discussing the challenges and direction when transforming from embedded systems to CPS. The enormous market size of the embedded systems has been demonstrated. Once again, CPS is described as embedded systems with networking capabilities. The authors present a good background literature with discussions on the co-simulation of the CPS and the distributed computing. Five major research directions in this field were listed as modeling CPS, verifying properties of CPS, designing for robust CPS, controlling the CPS, and ensuring secure and safe CPS.

Osswald et al. (2014) stated that one of the major challenges in automotive CPS is identified as the large interconnection of various devices in the system, which was unseen in the embedded systems. The large number leads to extensive number of possible states of the system and validating each state becomes impossible. Also, any violation by any device within the system may lead to a catastrophe, which requires critical care.

The challenges of trade-offs in the CPS were discussed by Anderson and Culler (2014). The terms embedded systems and CPS are used interchangeably to discuss various trade-offs in energy budget, storage, radio, flash, power, inter-module compatibility, and supervisory control.

6.4 Security of Embedded Systems in CPS

The energy sector has been greatly influenced by the advancement of technology in the field of CPS. The novel domain of smart grids is a result of integrating computers for remote control and networking in managing the core components of a power grid. Raciti and Nadjm-Tehrani (2013) presented a work on smart grid concerning on security aspect. The authors present security issues in the advanced metering infrastructures (AMI) of the smart grids and propose an anomaly detector based on trust platform. They argued that despite having a heavy security level assigned to the AMIs, there is still room for security breaches and the proposed technique can help improve the security.

Embedded systems are generally application specific, immensely optimized, and highly cost-effective systems. These systems consist only those components that are required for the application, thus they have various constraints which make considerations to combine security features into them difficult. Some of the constraints are cost sensitivity, energy usage, development environment, peripheral and processing power, etc. (Koopman 2004). Despite the above mentioned constraints, it is very essential for every embedded system to include at least the minimal amount of security, since these systems are being implemented in all walks of life and a security attack may lead to heavy losses, fatalities, and catastrophes.

Ricci and McGinnes (2003), in 2003, defined a five-step technique that can ensure better security of the embedded systems. These steps are to identify the threat, set the targets for security, assess the risks of security failure, devise countermeasures to overcome the threats, and finally assure that the countermeasures remain effective. The authors also give few recommendations toward embedded system security. One of the most elaborate studies on the security aspects of the embedded systems has been provided by Grand in 2004 (Grand 2004). It lists the security principles provided by Stoneburner, Hayden, and Feringa on behalf of National Institute of Standards and Technology (NIST) in 2001 (Stoneburner et al.

2001), which was revised in 2004 (Stoneburner et al. 2004). Grand gives the classification of the security based on the attack, attacker, threat, difficulty, product accessibility, and threat. Just like Ricci and McGinnes (2003), the first step toward the product design is identified as the risk assessment and management, which gives a broad idea on the security threats and the measures to be taken to overcome those threats. Some of the threats are interception (Eavesdropping), interruption (Denial-of-service (DoS) attack, malicious destruction of a hardware device, intentional erasure of program or data contents), modification, fabrication, competition (or Cloning), theft of service, spoofing, privilege escalation, etc. Various solutions to these have also been discussed for product enclosure, the board and PCB, and the firmware. In 2006, Grand (2006) discusses how hacking could be used to dismantle, reverse engineer, and modify products to perform various functions like breaking the authentication and copyright or perform a task which was never intended to do by the designers.

Ravi et al. (2004) identified six requirements for an embedded system. They are user identification, secure network access, secure communication, secure storage, secure content, and availability. Some of the security techniques like ciphers and hashing algorithms have been discussed. Also the cryptographic techniques like security protocol, digital certificate, digital rights management, and secure storage and execution have been listed. Software attacks and their countermeasures in case of complexity, extensibility, and connectivity have also been described. Tampering attacks like physical attack, timing analysis, power analysis, fault induction, and electromagnetic analysis have also been taken into consideration to design a secure architecture for the embedded systems. The paper was further extended and presented in Ravi et al. (2004). Since embedded systems are physical devices with logical functionalities inside, the attacks on them are classified into physical attack and logical attack. It also discussed the challenges like the processing gap in the embedded devices, the battery gap, flexibility in tasks and application, tampers resistance, assurance gap, and the cost of the product, in broad detail and also examines the requirements.

Koopman et al. (2005) discussed an undergraduate course on embedded system at Carnegie Mellon University. The paper clearly states that although the research in security has been around for some time, the research on embedded system security has been very minimal. Security techniques like authentication and access control, data privacy and integrity, software security, and security policies have been discussed.

Since the inception of the CPS, in 2006, that builds on the foundations of embedded systems, the research has shifted from embedded systems to CPS and thus a very few papers are being published in this field. Parameswaran and Wolf (2008) described an overview of embedded systems security in 2008. The characteristics and the vulnerabilities of general embedded systems have been listed and explained. The software attack and side channel attack on these systems have been described with their countermeasures. Some of the countermeasures for side channel attacks are the masking, window method, dummy instruction insertion, code/algorithm modification and balancing. Miller and Schorcht (2010) provided an overview of the performance of various cryptographic techniques used in the embedded systems. It is a part of the research conducted in the Kaspersky Labs. The various algorithms like data encryption standard (DES), advanced encryption standard (AES), 3DES, Blowfish, Twofish, IDEA, CAST, FEAL, SEAL, RC4, etc., have also been tabulated. The comparisons were based on the correctness and reliability, duration of encryption, energy consumption, memory needs and safety. But the authors did not provide any recommendations or improvements to the embedded system security.

6.4.1 Transformation of Security

The embedded systems and the CPS are used interchangeably by majority of the authors around the journals. But there are certainly few differences among the traditional embedded systems and the CPSs. This section provides an understanding of the current interrelationship among the two concepts by reviewing through the various research works being published from the beginning to the current times.

Mueller (2006) in 2006 highlighted three major issues in CPSs. Security being one of the issues, he pointed its seriousness in power grid and critical infrastructure and recommended the creation of software simulation for the IEC 61850 standard implemented on smart grids. The terms embedded systems and CPSs were treated as the same concept by the author.

Lee (2008), in his paper on the design challenges, has raised questions on security issues, risks, mitigations, and techniques required for CPS as transforming from embedded systems. Cárdenas et al. (2008), in 2008, proposed a mathematical framework to analyze attacks in their paper addressing the research challenges of security of control systems. The research tries to answer two main questions proposed by the authors and, in the process, proposes a framework which can help detect and survive attacks. The causes for the vulnerabilities and threats for the CPS as compared to embedded systems have been listed. The causes are that the human controllers have been replaced by computers, the systems have been networked, IT solutions are done using commodity software or protocols, openness in design, increase in size and functionality, more number of IT professionals around the globe, and increased number of cybercrimes. The security effort is shifting from reliability to protection against malicious attacks as the technology shifts from embedded systems to CPS. The efforts are also taken by NERC, NIST, ISA, etc., and the combined goals are listed as creating awareness, and help operators to design security policy and recommend basic prevention methods. The new security issues like effects on estimation, control and the physical world in case of an attack have been discussed and few recommendations have been provided. The authors modeled these issues by use of mathematics and mainly discussing DoS and deception attacks. Though it is a theoretical work, it provided a good information and motivation to many other researchers.

In 2009, Sun et al. (2009) grouped embedded systems into three categories based on their application, complexity, and criticality. Among them, the moderately critical and highly critical embedded systems are termed as CPS. The authors discuss the security threats occurring due to the contradictions between the safety requirements and the security needs. To overcome this, the authors proposed a framework to detect the conflicts early during the designing stage. Sha et al. (2009) put main emphasis on medical CPS (also termed as real-time networked embedded systems) and their security. The authors suggest that there is a need to reexamine the existing hardware and software implement into CPSs. Along with various other aspects, security characteristics must be comprised in programming abstractions. Since the study of field of security is limited, the authors recommend building a prototypical CPS model and establish metrics that address errors, faults, and security attacks. The security gaps must then be verified and validated. The assumptions regarding the external environments need to be studied in the context of safety, robustness, and security.

In 2010, Rajkumar (Rajkumar et al. 2010) described the challenges of the revolution of CPS. The intersection of physical and cyber components needs the enforcement of security and privacy requirements, which would be achieved by implementing innovative solutions. The possible attacks are to be countered by developing security policy, intrusion detection, and mitigation. Safety and security of the CPSs from security attacks and errors are identified as one of the critical issues in CPS. The physical nature can be exploited by implementing the location, time, and tag-based mechanisms for security solutions. The programming of these systems must also include the security features embedded in them.

In their research on smart grids, Khurana et al. (2010) described the security issues connected with the smart grids. Smart grid forms one of the critical infrastructures in any developed country which makes their security a vital issue to address. Trust is identified an important component for security in CPS. Security and privacy in the communication between the devices were considered important. The security management in taking care of the public key infrastructure and updating the cryptographic keys and key generation were discussed. Some of the recent developments like new authentication techniques for transmission substations and SCADA networks, policy-based data sharing, and attestation had been listed and explained.

Akella et al. (2010) worked on the security of the information flow in CPSs and presented a model for its analysis. The security of information flow is related to confidentiality of the data, the aspect of security that the authors claimed to be neglected by most researchers. The information flowing are the interactions that take place between the cyber and the physical parts of the CPS. The Bell–LaPadula (Bell and LaPadula 1973; Bell 2011) model and other related models were discussed with their limitations and their approach was defined using process algebra concept. The authors took the applications of pipeline system and electric power grid to demonstrate the approach. Although the paper is dedicated to security of CPS, in both the applications, the embedded systems were included as an integral part of CPSs.

Wang et al. (2011), in 2011, proposed a CPeSC3 architecture to obtain a secure medical CPS. The architecture comprised components like sensing, communication, computation and resource scheduling, cloud and security core. The communication and the computing cores basically consist of embedded systems which are integrated to form the medical CPSs. The CPeSC3 model uses and combines concepts and models from cloud computing, security and real-time systems. The security core of the model implements various security measures like symmetric, asymmetric and hybrid keys, error recovery for memory devices, data redundancy, and cloud backup. For the healthcare CPS, home, office and hospital environment were considered and the security was analyzed and planned.

Sanislav and Miclea (2012) conducted a survey on CPS and its challenges in 2012. Security is considered as one of the major challenges for CPS and the factors influencing it are usability, management, and adaptability. In contrast to the embedded systems, CPS is connected more directly to the physical world and processes, this leads to a vital challenge of detecting and adapting to environmental changes and attacks. Shafi (2012) focused on the security requirements, objectives, threats, attacks, and solutions in CPS. He listed the security objectives as confidentiality, integrity, availability, and authenticity, which are similar to that of the embedded systems. Various attacks, concerns, solutions, and application-specific security issues were also discussed. Yet another publication on the challenges of CPS was done by Conti et al. (2012) in the same year. One of the major challenges was to devise a security and privacy of the small, inexpensive and resource-constraint embedded systems implemented in CPSs. Many coinciding concepts of embedded systems and CPSs related to information management were discussed. Some of the features are the data storage, scalability of data handling, complexity of data, flexibility, reproducibility of the computations, and the quality of information. The embedded systems possess the computational, sensing and actuating capabilities and the cyberization leads to CPS. With the advancement of technology and sophistication, the malware producers and attackers have also advanced to great extents. The authors claim that malwares are now observed in devices and systems that were earlier claimed to have been unreachable by attacks. This requires solutions from three classes, namely secure hardware, pure software, and hybrid technology.

Broy et al. (2012), working on a German project on CPS, discussed some challenges involved in its advancement. Among the immediate and major challenges is that of the safety, security, and privacy. As the embedded systems become more capable and networked, their security becomes more vital and vulnerable. Confidentiality, integrity, and authenticity are identified as the three basic cores of the security for CPS. Other aspects related to security are dependability, reliability, availability, and maintainability. The "safety@runtime" and "security@runtime" are the two requirements that require attention as there is no technique developed yet to conduct uninterrupted maintenance and servicing of CPS. One of the recommendations provided is to consider security during both the development stage and operational stage of CPS. The requirements listed are security hardware, secure environment, new procedures for trust, security management and engineering, stringent test and analysis, safety/security regulations, standardized development and mechanisms.

In 2013, Serpanos and Voyiatzis (2013) discussed various security challenges concerning the embedded systems. Examining the embedded systems in CPS, encryption and authentication were the two mechanisms identified to provide security. For encryption, elliptical curve cryptography is recommended over the public key cryptography, as the former is computationally less demanding than the latter. Key management is considered the most critical part of the encryption system, since the leakage of keys makes even the toughest cryptographic system pliable. As the attackers' techniques and methods advance with time, the designers are recommended to upgrade the systems regularly. Since, once operational, availability of the system becomes compulsory, the upgrades must always be done during runtime. The remote management is suggested to be equipped with powerful intrusion detection systems to prevent attacks and access to critical information. The general DoS targeting specific hardware or software and the DDoS that collapses the whole system are also a concern for embedded systems in CPS.

On their paper on privacy in CPS, Petroulakis et al. (2013) identified that one of the vital issues in embedded systems is that a very small amount of research has been done on the designing of computer-based control systems. They argued that embedded systems can easily reveal information about the location, routing algorithm, or any other private sensitive element and since CPS are deployed in harsh and uncontrollable environment, attacks could be effortless and obvious. They proposed a privacy-level model that generates privacy countermeasures to overcome the attacks involved. Using this model, the operator of the system would be able to assign various privacy levels to the system based on the privacy requirements of the network. Trapp et al. (2013) provided a roadmap for the safety and security of the CPSs in their paper. Some of the issues for the embedded systems transforming into CPS, as pointed out by them, are the development of the ability to dynamically adapt to the changing context of runtime and environment and to follow strict rules and rigorous safety assurance case. These both issues are mainly linked to the safety and security of the CPS. The authors conducted an extensive study on the various safety aspects and measures to ensure safety of the CPS. A common safety interface framework based on SafetyCertification@Runtime was found to be promising to ensure runtime safety, verification, and validation.

The paper by Greenwood et al. (2015), that discusses the evolution of embedded systems to CPS, presents one of the latest researches in this field. The authors believe that the Neo-Darwinistic method known as evolvable and adaptive hardware would lead to evolution of the systems for better solutions and applications. A list of five major properties was identified for an ideal CPS. Security and safety are two among those five properties. Thus, they are given critical importance and require further investigation and innovation in this direction. The current embedded system design focuses more on the cyber aspect and when designing CPS, this attitude requires a change. The focus must also be set over the hardware, communication, and computation with security attached in each of this section.

The paper by Ye (2015) proposed a security protection design for CPS. CPS is a technology that embeds the communication and computation capabilities into the traditional embedded systems. Security is a major issue and the paper gives an elaborate discussion on it with examples of attacks on nuclear power plant, automobiles, medical industry, military, etc. Among the security measures, the "TrustZone" created by ARM was discussed. It sets a separated safety zone in the processor to prevent network attacks on the embedded system. Another protection method is creating three subdivisions called engineer station, controller, and APC server within the CPS. Each implements its security measures in isolation by examining, tracking, and protecting the system. This technique is generally implemented in industrial control systems. The bug in the embedded system development also needs to be taken care of while its implementation in CPS.

In the systematic literature review exclusively on the security of CPS by Dong et al. (2015), they discussed the various aspects and developments in this field till date. The safety of the systems was given more priority over the security. The security requirements of a general CPS are identified as confidentiality, integrity, and availability. The latest issues are listed as security protocol seamlessness, global trust management, and privacy protection. The aspect of risk management in CPS forms a vital part of CPS security and must be researched further to find better solutions. The security in various layers of CPS was also discussed.

6.5 Smart Grids Security

Sridhar et al. (2012) presented a paper on the security of the electric power grids. The paper discussed the importance of trust, risk evaluation, and cybersecurity in CPS. The embedded systems are used all over the electric grid systems to support, monitor, and control the functions within and their protection falls under the device security sector of the CPS. The remote attestation is an important part of the smart grid technology. Some of the devices are smart meter and intelligent electronic device (IED). These devices are implemented using embedded systems and require ample security to protect it from tampering. A cryptographic sign loaded firmware for the purpose was discussed (LeMay and Gunter 2012), but its limitations were also pointed out. The security issues must be addressed very carefully to ensure integrity of these devices and many recommendations like risk modeling and mitigation and trust management, etc., had been provided by the authors to achieve the security.

Liu et al. (2012) researched on the same field to investigate the security and privacy issues in the smart grids. The security issues within the devices like programmable logic controllers (PLCs), remote terminal unit (RTUs), and IEDs in the smart grids arise from the embedded systems. Some security standards and trade-offs were discussed. One of the cryptography method mentioned was the

AES-CCM 128-bit shared key for encryption of data. For the purpose of malware attack protection, they discuss the procedure proposed by Metke and Ekl (2010). The three procedures described are upgrade patch that can be delivered by issuing public key to each device, hot assurance boot, and all mobile code must be strictly controlled between supplier and operators.

A survey was presented by Wang and Lu (2013) discussing the challenges in cybersecurity of smart grids. Millions of embedded systems with computational limitations combined together to form the smart grid. The computation efficiency thus becomes a critical issue to enable encryption scheme in the smart grid. Although asymmetric key cryptography is stronger than the symmetric, it requires more computations and thus its implementation becomes limited for the embedded systems. To overcome this, the recommendation is to either increase the hardware support or make central processing unit (CPU) more power. Authentication is yet another security measure and it requires high efficiency, tolerance against attacks and fault, and the support to multicast technique. The DES-CBC and the RSA key schemes were also analyzed for the same. RSA is still not a feasible choice due to the computational inefficiency. The three kinds of multicast authentication listed and compared were secret information, asymmetry, time asymmetry, and hybrid asymmetry. Several key management systems were also discussed in the survey.

One of the latest publications in this field was by Zimmer et al. (2015), which proposed and discussed methods for intrusion detection in the controllers of power grids. The main focus of the authors was the real-time embedded systems in the power grids that are bound to intrusion attacks. Using the micro-timings of the systems, the code execution progress is tracked and the possible attacks can be detected. The intrusion is detected using two ways, by self-checking and by OS scheduler. This helps detect execution of unauthorized code. The method developed consisted of three mechanisms named as t-Rex, T-ProT, and T-AxT. The mechanisms were tested on both simulation and embedded system hardware and found to have been a novel approach for the intrusion detection.

6.6   Gaps and Limitations

6.6.1   Complexity

The embedded systems market is about 850 billion Euros worldwide (Union 2016). This shows the great impact it has on the world in the current times. The hardware being manufactured is getting more powerful every day. Also, the networking of the embedded systems with desired OSs has become the need of the hour (Herausforderung and Wirtschaft 2009). This is giving rise to demand in embedded systems with complex requirements like higher capabilities, lower power consumption, longer durability, heavier memory, better networking options, and faster processing. Billions of networked and embedded systems are expected to be connected to provide a fully networked and functional IoT (Herausforderung and Wirtschaft 2009). The European Union (EU) has realized its importance and initiates the Horizon 2020, an investment of about 140 million Euros on embedded systems and CPSs (Union 2016). The goal is to obtain higher reliability, dependability, autonomy, reduced power consumption, more connectivity, mixed criticality, and new business models (Haydn 2013).

The second major complexity is that of the software for the embedded and CPSs (Sprinkle 2008). Software is equally important as the hardware and most researchers do not realize this fact (Sullivan and Krikeles 2008). The software governs how the hardware is handled based on the requirements. Software is used to receive the data and process them to control and execute the expected tasks. These can be the application specific software codes and/or the OSs. As the systems become more complex and the network more diverse, the software development becomes more complex and extensive. The software must also aim to make the system more secure and safe by protecting the confidentiality, integrity, and authenticity of the system, its tasks and data (Tripakis and Sengupta 2014; Zhang et al. 2012).

Error in both, hardware or software may lead to heavy losses and complications. The current engineering and technology must take care of this fact (Mueller 2006). The errors are not only needed to be diagnosed but also an effort is required to eliminate them during the earliest possible phase of designing and fabrication. Time to produce and time to market is yet another important complexity which forms a vital business strategy in this field. Marketing a product too early or too late may lead to drastic losses.

6.6.2 Interconnecting the Internet and Embedded Systems

A CPS is incomplete without the functionality of Internet. This requires the embedded systems to develop Internet capabilities. The Transmission Control Protocol/internet protocol (IP) protocols are being implemented in the embedded systems to achieve this functionality. The embedded systems also require including modules like the wireless connectivity and the Ethernet controller. Since most of these systems are battery powered, it becomes compulsory for the systems to be efficient in the power usage and avoid draining the available power. Thus, one of the challenges in these systems is to include an Internet connectivity which not only serves the purpose but also does not add a burden on the power management. Lightweight IP (LwIP) is one of the versions of TCP/IP which is being implemented for this purpose (Dunkels 2001, 2003). The connectivity will enable the objective of achieving the IoT.

6.6.3   Trust Establishment

Throughout this article, the importance of security has been highlighted. Trust and reputation can form one of the most effective techniques toward security of the CPSs. The security aspect of CPSs has been studied by many researchers (Wang et al. 2011; Kirkpatrick et al. 2009; Wu et al. 2011; Zimmer et al. 2010; Zhu et al. 2011b; Poolsappasit et al. 2012; Sommestad et al. 2009). The trust management is now gaining attention and has been studied extensively in context of wireless sensor networks (WSNs) and IoT (Singh et al. 2014; Che et al. 2015; Yan et al. 2014; Reshmi and Sajitha 2014). The authors of this article are working on the trust management in CPS. Ali et al. (2015) and Ali and Anwar (2012) have proposed a two-tier trust system that can maintain the security of the CPS. The authors are working on this trust management technique and trying to bring improvements into it.

6.6.4   Trade-Offs

Trade-off is a difficult aspect of the CPS systems and requires very deep and strict understanding of the requirements of a given system so as to make the best possible balance between various factors governing the efficient functioning of the system.

Stringent security measure leads to more computations in the embedded systems of the CPS. This leads to a decrease in the efficiency of the implementation of the CPS. Thus, a trade-off between the security and the efficiency is an aspect that must be taken care of, as shown in Liu et al. (2012). Privacy is a crucial aspect of security and protecting it is a very vital part of securing the CPS. But creating various privacy levels for the different layers of the CPS leads to energy consumption by the embedded systems. This is due to the fact that the amount of computations needed to protect the privacy increases with the increment of the level assigned for privacy. Since the embedded systems in CPS are generally bound to constraints in the power supply and energy consumption, the trade-off between security and energy consumption is very important to CPS development. This has been demonstrated and proved by Petroulakis et al. (2013).

Generally, longer key size leads to stronger security of the systems. But this also leads to more time consumption and performance delay in the physically constrained embedded systems. The trade-off between security and latency thus requires a comprehensive study and investigation, as provided by Wang and Lu (2013). Zimmer et al. (2015) also attempted to provide better protection against attacks by observing the relationship between security and timeliness of the CPS. The embedded systems that require real-time responses and execution tasks may take longer duration to perform a task when the security mechanism is made stronger with more security features. Thus studying the trade-off among the security strength and the timeliness in real-time CPS applications requires experimentation and analysis.

Upgrading the systems and software of the CPS is an important part of fighting the attackers and also improving the functioning and implementation. But upgrading may also lead to security risks, like mistakenly replacing good software by a malicious one or a good hardware by a suspicious one. Defenses thus must be applied on both the communication protocols and the physical layer. Thus, there exists a trade-off between upgrading to an advanced software/hardware and the protection of the security of the CPS. This has been demonstrated by Serpanos and Voyiatzis (2013).

6.7   Conclusion

Embedded systems are processing units that are connected larger systems through sensors, actuators, controllers, and communication peripherals (Beetz and Böhm 2012). It has flourished into every walk of human life and this can proved by the fact that over 90% of the processors produced today are used in embedded systems (Intel 2016; Twente 2016). The applications are found in field like health care, transportation, intelligence, communication, control, governance, etc. With the introduction of Internet, the embedded systems gained further reach to form the CPSs, coined by Helen Gill in 2006 (Gunes et al. 2014). The transformation from embedded systems to CPSs has led to various advancements in the modeling, design, programming tools, languages, standards, and security which helped expand its applications.

References

Akella, R., Tang, H., & McMillin, B. M. (2010). Analysis of information flow security in cyber–physical systems. *International Journal of Critical Infrastructure Protection, 3,* 157–173.

Ali, S., & Anwar, R. W. (2012). Trust based secure cyber physical systems. In *Workshop Proceedings: Trustworthy Cyber-Physical Systems*, Computing Science, Newcastle University, 2012.

Ali, S., Anwar, R. W., & Hussain, O. K. (2015). Cyber security for cyber-physical systems: A trust based approach. *Journal of Theoretical and Applied Information Technology, 71,* 144–152.

Alippi, C. (2014). *Intelligence for embedded systems*. Springer.

Andersen, M. P., & Culler, D. E. (2014). System design trade-offs in a next-generation embedded wireless platform.

Ashok, A., Hahn, A., & Govindarasu, M. (2014). Cyber-physical systems of wide-area monitoring, protection and control in a smart grid environment. *Journal of Advance Research, 5,* 481–489.

Backhaus, S., Bent, R., Bono, J., Lee, R., Tracey, B., Wolpert, D., et al. (2013). Cyber-physical security: A game theory model of humans interacting over control systems. *IEEE Transactions on Smart Grid, 4,* 2320–2327.

Baker, S. A., Waterman, S., & Ivanov, G. (2009). *In the crossfire: Critical infrastructure in the age of cyber war.* McAfee, Incorporated.

Bartocci, E., Hoeftberger, O., & Grosu, R. (2014). Cyber-physical systems: Theoretical and practical challenges. *ERCIM News, 2014.*

Battram, P., Kaiser, B., & Weber, R. (2015). A modular safety assurance method considering multi-aspect contracts during cyber physical system design.

Beetz, K., & Böhm, W. (2012). Challenges in engineering for software-intensive embedded systems. In K. Pohl, H. Hönninger, R. Achatz, & M. Broy (Eds.), *Model-based engineering of embedded systems.* Berlin, Heidelberg: Springer.

Bell, D. E., & LaPadula, L. J. (1973). *Secure computer systems: Mathematical foundations.* DTIC Document.

Bonakdarpour, B. (2008). Challenges in transformation of existing real-time embedded systems to cyber-physical systems. *ACM SIGBED Review, 5,* 11.

Broy, M. (2013). Engineering cyber-physical systems: Challenges and foundations. In *Complex systems design & management.* Springer.

Broy, M., Cengarle, M. V., & Geisberger, E. (2012). Cyber-physical systems: Imminent challenges. In *Large-scale complex IT systems. Development, operation and management.* Springer.

Broy, M., & Schmidt, A. (2014). Challenges in engineering cyber-physical systems. *Computer,* 70–72.

Bujorianu, M. L., & Mackay, R. S. (2014). Complex systems techniques for cyber-physical systems: Position paper. In *Proceedings of the 4th ACM SIGBED International Workshop on Design, Modeling, and Evaluation of Cyber-Physical Systems* (pp. 27–30). ACM.

Cárdenas, A. A., Amin, S., & Sastry, S. (2008). Research challenges for the security of control systems. In *HotSec.*

Casale-Rossi, M., De Micheli, G., Bagherli, J., Collette, T., Domic, A., Symanzik, H., et al. (2015). The future of electronics, semiconductors, and design in Europe: Panel. In *Proceedings of the 2015 Design, Automation & Test in Europe Conference & Exhibition* (pp. 1726–1728). EDA Consortium.

Che, S., Feng, R., Liang, X., & Wang, X. (2015). A lightweight trust management based on Bayesian and Entropy for wireless sensor networks. *Security and Communication Networks, 8,* 168–175.

Conti, J. (2010). The day the samba stopped [power blackouts]. *Engineering & Technology, 5,* 46–47.

Conti, M., Das, S. K., Bisdikian, C., Kumar, M., Ni, L. M., Passarella, A., et al. (2012). Looking ahead in pervasive computing: Challenges and opportunities in the era of cyber-physical convergence. *Pervasive and Mobile Computing, 8,* 2–21.

Das, S. K., Kant, K., & Zhang, N. (2012). *Handbook on securing cyber-physical critical infrastructure.* Elsevier.

Davies, N., & Gellersen, H.-W. (2002). Beyond prototypes: Challenges in deploying ubiquitous systems. *Pervasive Computing, IEEE, 1,* 26–35.

Denning, D. E. (2000). Cyberterrorism: The logic bomb versus the truck bomb. *Global Dialogue, 2,* 29.

Dong, P., Han, Y., Guo, X., & Xie, F. (2015). A systematic review of studies on cyber physical system security. *International Journal of Security and Its Applications, 9,* 155–164.

Dunkels, A. (2001). Design and implementation of the lwIP TCP/IP stack. *Swedish Institute of Computer Science, 2,* 77.

Dunkels, A. (2003). Full TCP/IP for 8-bit architectures. In: *Proceedings of the 1st International Conference on Mobile Systems, Applications and Services* (pp. 85–98). ACM.

Eidson, J. C., Lee, E. A., Matic, S., Seshia, S. A., & Zou, J. (2012). Distributed real-time software for cyber–physical systems. *Proceedings of the IEEE, 100,* 45–59.

Elliott Bell, D. (2011). Bell–La Padula model. In *Encyclopedia of cryptography and security* (pp. 74–79).

Farwell, J. P., & Rohozinski, R. (2011). Stuxnet and the future of cyber war. *Survival, 53,* 23–40.

Fink, J., Ribeiro, A., & Kumar, V. (2012). Robust control for mobility and wireless communication in cyber–physical systems with application to robot teams. *Proceedings of the IEEE, 100*, 164–178.

Fitzgerald, J., Larsen, P. G., & Verhoef, M. (2014). From embedded to cyber-physical systems: Challenges and future directions. In *Collaborative design for embedded systems*. Springer.

Grand, J. (2004). Practical secure hardware design for embedded systems. In *Proceedings of the 2004 Embedded Systems Conference*.

Grand, J. (2006). Research lessons from hardware hacking. *Communications of the ACM, 49*, 44–49.

Greenberg, A. (2008). Hackers cut cities' power. In *Forbes, Jaunuary*.

Greenwood, G., Gallagher, J., & Matson, E. (2015). Cyber-physical systems: The next generation of evolvable hardware research and applications. In *Proceedings of the 18th Asia Pacific Symposium on Intelligent and Evolutionary Systems* (Vol. 1, pp. 285–296). Springer.

Gunes, V., Peter, S., Givargis, T., & Vahid, F. (2014). A survey on concepts, applications, and challenges in cyber-physical systems.

Gupta, A., Kumar, M., Hansel, S., & Saini, A. K. (2013). Future of all technologies-the cloud and cyber physical systems. *International Journal of Enhanced Research in Science Technology and Engineering, 2*.

Gurgen, L., Gunalp, O., Benazzouz, Y., & Gallissot, M. (2013). Self-aware cyber-physical systems and applications in smart buildings and cities. In *Proceedings of the Conference on Design, Automation and Test in Europe* (pp. 1149–1154). EDA Consortium.

Hall, E. C. (1996). *Journey to the moon: The history of the Apollo guidance computer*. Aiaa.

Halperin, D., Heydt-Benjamin, T. S., Ransford, B., Clark, S. S., Defend, B., Morgan, W., et al. (2008). Pacemakers and implantable cardiac defibrillators: Software radio attacks and zero-power defenses. In *IEEE Symposium on Security and Privacy, SP 2008* (pp. 129–142). IEEE.

Haydn, T. (2013). *Cyber-physical systems: Uplifting Europe's innovation capacity*. Brussels: Communications Networks, Content & Technology Directorate-General.

Herausforderung, & Wirtschaft, C. F. D. D. (2009). "Embedded Software"—Challenge and opportunities for the German economy. Bundesministerium für Bildung und Forschung (Federal Ministry of Education and Research).

INTEL. (2016). *Introduction to embedded systems* [Online]. Available: http://www.intel.com/education/highered/Embedded/Syllabus/Embedded_syllabus.pdf. Accessed January 01, 2016.

Jalali, S. (2009). Trends and implications in embedded systems development. *TCS white paper*.

Johnson, T. T., Bak, S., & Drager, S. (2015). Cyber-physical specification mismatch identification with dynamic analysis. In *Proceedings of the ACM/IEEE Sixth International Conference on Cyber-Physical Systems* (pp. 208–217). ACM.

Kamal, R. (2008). *Embedded systems 2E*. Tata McGraw-Hill Education.

Karnouskos, S., Colombo, A. W., & Bangemann, T. (2014). Trends and challenges for cloud-based industrial cyber-physical systems. In *Industrial cloud-based cyber-physical systems*. Springer.

Keller, I., Lehmann, A., Franke, M., & Schlegel, T. (2014). Towards an interaction concept for efficient control of cyber-physical systems. In *Virtual, augmented and mixed reality. Designing and developing virtual and augmented environments*. Springer.

Khurana, H., Hadley, M., Lu, N., & Frincke, D. A. (2010). Smart-grid security issues. *IEEE Security & Privacy*, 81–85.

Kim, K.-D., & Kumar, P. R. (2012). Cyber–physical systems: A perspective at the centennial. *Proceedings of the IEEE, 100*, 1287–1308.

Kirkpatrick, M., Bertino, E., & Sheldon, F. T. (2009). Restricted authentication and encryption for cyber-physical systems. In *DHS CPS Workshop Restricted Authentication and Encryption for Cyber-physical Systems*. Citeseer.

Koopman, P. (2004). Embedded system security. *Computer, 37*, 95–97.

Koopman, P., Choset, H., Gandhi, R., Krogh, B., Marculescu, D., Narasimhan, P., et al. (2005). Undergraduate embedded system education at Carnegie Mellon. *ACM Transactions on Embedded Computing Systems (TECS), 4,* 500–528.

Koubâa, A., & Andersson, B. (2009). *A vision of cyber-physical internet.* Portugal: Polytechnic Institute of Porto.

Lee, E. (2015a). *Ptolemy II* [Online]. Available: http://ptolemy.eecs.berkeley.edu/ptolemyII/. Accessed July 14, 2015.

Lee, E. (2015b). *The Ptolemy project* [Online]. Available: http://ptolemy.eecs.berkeley.edu/. Accessed July 14, 2015.

Lee, E. A. (2006). Cyber-physical systems-are computing foundations adequate. In *Position Paper for NSF Workshop on Cyber-Physical Systems: Research Motivation, Techniques and Roadmap.*

Lee, E. A. (2007). Computing foundations and practice for cyber-physical systems: A preliminary report. *Technical Report UCB/EECS-2007-72.* Berkeley: University of California.

Lee, E. A. (2008). Cyber physical systems: Design challenges. In *11th IEEE International Symposium on Object Oriented Real-Time Distributed Computing (ISORC)* (pp. 363–369). IEEE.

Lee, E. A. (2010). CPS foundations. In *Proceedings of the 47th Design Automation Conference* (pp. 737–742). ACM.

Lee, E. A. (2015c). The past, present and future of cyber-physical systems: A focus on models. *Sensors, 15,* 4837–4869.

Lee, E. A., & Seshia, S. A. (2014). *Introduction to embedded systems—A cyber-physical systems approach.* LeeShehia.org.

Lee, I., Pappas, G. J., Cleaveland, R., Hatcliff, J., Krogh, B. H., Lee, P., et al. (2006). High-confidence medical device software and systems. *Computer, 39,* 33–38.

Lee, I., & Sokolsky, O. (2010). Medical cyber physical systems. In *Proceedings of the 47th Design Automation Conference.* Anaheim, California: ACM.

Lemay, M., & Gunter, C. (2012). Cumulative attestation kernels for embedded systems. *IEEE Transactions on Smart Grid, 3,* 744–760.

Leyden, J. (2008). Polish teen derails tram after hacking train network. *The Register* (Vol. 11).

Liu, J., Xiao, Y., Li, S., Liang, W., & Chen, C. (2012). Cyber security and privacy issues in smart grids. *Communications Surveys & Tutorials, IEEE, 14,* 981–997.

Ma, Z., Marchal, P., Scarpazza, D. P., Yang, P., Wong, C., Gómez, J. I., et al. (2007). *Systematic methodology for real-time cost-effective mapping of dynamic concurrent task-based systems on heterogenous platforms.* Springer Science & Business Media.

Magureanu, G., Gavrilescu, M., & Pescaru, D. (2013). Validation of static properties in unified modeling language models for cyber physical systems. *Journal of Zhejiang University Science C, 14,* 332–346.

Marwedel, P. (2010). *Embedded system design: Embedded systems foundations of cyber-physical systems.* Springer Science & Business Media.

Metke, A. R., & Ekl, R. L. (2010). Security technology for smart grid networks. *IEEE Transactions on Smart Grid, 1,* 99–107.

Miller, A., & Schorcht, G. (2010). Embedded systems security: Performance investigation of various cryptographic techniques in embedded systems.

Miller, W. B. (2014). Classifying and cataloging cyber-security incidents within cyber-physical systems.

Mitchell, R., & Chen, I.-R. (2015). Behavior rule specification-based intrusion detection for safety critical medical cyber physical systems. *IEEE Transactions on Dependable and Secure Computing, 12,* 16–30.

Molina, J. M., Damm, M., Haase, J., Holleis, E., & Grimm, C. (2014). Model based design of distributed embedded cyber physical systems. In *Models, methods, and tools for complex chip design.* Springer.

Mosterman, P. J., & Zander, J. (2015). Cyber-physical systems challenges: A needs analysis for collaborating embedded software systems. In *Software & systems modeling* (pp. 1–12).

Mueller, F. (2006). Challenges for cyber-physical systems: Security, timing analysis and soft error protection. In *High-Confidence Software Platforms for Cyber-Physical Systems (HCSP-CPS) Workshop* (p. 4), Alexandria, Virginia.

Nath, P. K., & Datta, D. (2014). Multi-objective hardware–software partitioning of embedded systems: A case study of JPEG encoder. *Applied Soft Computing, 15,* 30–41.

Navet, N., & Simonot-Lion, F. (2008). *Automotive embedded systems handbook.* CRC Press.

Osswald, S., Matz, S., & Lienkamp, M. (2014). Prototyping automotive cyber-physical systems. In *Proceedings of the 6th International Conference on Automotive User Interfaces and Interactive Vehicular Applications* (pp. 1–6). ACM.

Parameswaran, S., & Wolf, T. (2008). Embedded systems security—An overview. *Design Automation for Embedded Systems, 12,* 173–183.

Parolini, L., Sinopoli, B., Krogh, B. H., & Wang, Z. (2012). A cyber–physical systems approach to data center modeling and control for energy efficiency. *Proceedings of the IEEE, 100,* 254–268.

Parvin, S., Hussain, F. K., Hussain, O. K., Thein, T., & Park, J. S. (2013). Multi-cyber framework for availability enhancement of cyber physical systems. *Computing, 95,* 927–948.

Petroulakis, N. E., Askoxylakis, I. G., Traganitis, A., & Spanoudakis, G. (2013). A privacy-level model of user-centric cyber-physical systems. In *Human aspects of information security, privacy, and trust.* Springer.

Pike, L., Sharp, J., Tullsen, M., Hickey, P. C., & Bielman, J. (2015). Securing the automobile: A comprehensive approach.

PLC, A. H. (2014). *Shaping the connected world.* Strategic Report 2014.

Pohlmann, U., Dziwok, S., Meyer, M., Tichy, M., & Thiele, S. (2014). A modelica coordination pattern library for cyber-physical systems. In *Proceedings of the 7th International ICST Conference on Simulation Tools and Techniques* (pp. 76–85). ICST (Institute for Computer Sciences, Social-Informatics and Telecommunications Engineering).

Poolsappasit, N., Dewri, R., & Ray, I. (2012). Dynamic security risk management using bayesian attack graphs. *IEEE Transactions on Dependable and Secure Computing, 9,* 61–74.

Ptolemaeus, C. (2014). *System design, modeling, and simulation: Using Ptolemy II.* Ptolemy.org. Berkeley, CA, USA.

Quinn-Judge, P. (2002). Cracks in the system. *TIME Magazine,* January 9, 2002.

Raciti, M., & Nadjm-Tehrani, S. (2013). Embedded cyber-physical anomaly detection in smart meters. In *Critical information infrastructures security.* Springer.

Rajkumar, R. R., Lee, I., Sha, L., & Stankovic, J. (2010). Cyber-physical systems: The next computing revolution. In *Proceedings of the 47th Design Automation Conference* (pp. 731–736). ACM.

Ravi, S., Raghunathan, A., Kocher, P., & Hattangady, S. (2004). Security in embedded systems: Design challenges. *ACM Transactions on Embedded Computing Systems (TECS), 3,* 461–491.

Reshmi, V., & Sajitha, M. (2014). A survey on trust management in wireless sensor networks. *International Journal of Computer Science & Engineering Technology, 5,* 104–109.

Ricci, L., & McGinnes, L. (2003). Embedded system security-designing secure system with windows CE. *Embedded Computer System,* 1–33.

Sanislav, T., & Miclea, L. (2012). Cyber-physical systems-concept, challenges and research areas. *Journal of Control Engineering and Applied Informatics, 14,* 28–33.

Segovia, F., Serrano, R., Górriz, J., Ramírez, J., & González, J. (2012). A DSP embedded system. Application to digital communication systems. In *Technologies Applied to Electronics Teaching (TAEE),* 2012 (pp. 196–200). IEEE.

Serpanos, D. N., & Voyiatzis, A. G. (2013). Security challenges in embedded systems. *ACM Transactions on Embedded Computing Systems (TECS), 12,* 66.

Sha, L., Gopalakrishnan, S., Liu, X., & Wang, Q. (2009). Cyber-physical systems: A new frontier. In *Machine learning in cyber trust.* Springer.

Shafi, Q. (2012). Cyber physical systems security: A brief survey. In *ICCSA Workshops* (pp. 146–150).

Sharp, J. A. (1986). An introduction to distributed and parallel processing.

Shi, J., Wan, J., Yan, H., & Suo, H. (2011). A survey of cyber-physical systems. In *International Conference on Wireless Communications and Signal Processing (WCSP)* (pp. 1–6). IEEE.

Shukla, S. K. (2015). Editorial: Schizoid design for critical embedded systems. *ACM Transactions on Embedded Computing Systems (TECS), 14*, 40e.

Sifakis, J. (2011). A vision for computer science—The system perspective. *Central European Journal of Computer Science, 1,* 108–116.

Singh, M., Sardar, A. R., Sahoo, R. R., Majumder, K., Ray, S., & Sarkar, S. K. (2014). Lightweight trust model for clustered WSN. In *Proceedings of the 3rd International Conference on Frontiers of Intelligent Computing: Theory and Applications (FICTA) 2014, 2015* (pp. 765–773). Springer.

Slay, J., & Miller, M. (2008). *Lessons learned from the maroochy water breach.* Springer.

Sommestad, T., Ekstedt, M., & Johnson, P. (2009). Cyber security risks assessment with bayesian defense graphs and architectural models. In *42nd Hawaii International Conference on System Sciences, HICSS'09* (pp. 1–10). IEEE.

Sprinkle, J. (2008). Grand challenges education and cross-cutting challenges in cyber-physical systems. In *National Workshop for Research on High-Confidence Transportation Cyber-Physical Systems: Automotive, Aviation & Rail*, Vienna, Virginia, USA.

Sridhar, S., Hahn, A., & Govindarasu, M. (2012). Cyber–physical system security for the electric power grid. *Proceedings of the IEEE, 100,* 210–224.

Stojmenovic, I. (2014). Machine-to-machine communications with in-network data aggregation, processing, and actuation for large-scale cyber-physical systems. *IEEE Internet of Things Journal, 1,* 122–128.

Stoneburner, G., Hayden, C., & Feringa, A. (2001). *Engineering principles for information technology security (a baseline for achieving security).* DTIC Document.

Stoneburner, G., Hayden, C., & Feringa, A. (2004). *Engineering principles for information technology security (a baseline for achieving security)*, Revision A. DTIC Document.

Sullivan, G., & Krikeles, B. (2008). Grand challenges for transportation cyber-physical systems. In *National Workshop for Research on High-Confidence Transportation Cyber-Physical Systems: Automotive, Aviation & Rail.*

Sun, M., Mohan, S., Sha, L., & Gunter, C. (2009). Addressing safety and security contradictions in cyber-physical systems. In *Proceedings of the 1st Workshop on Future Directions in Cyber-Physical Systems Security (CPSSW'09).*

Talbot, D. (2012). Computer viruses are "rampant" on medical devices in hospitals. *MIT Technology Review, 17,* 19.

Tan, Y., Goddard, S., & Perez, L. C. (2008). A prototype architecture for cyber-physical systems. *ACM Sigbed Review, 5,* 26.

Tham, C.-K., & Luo, T. (2013). Sensing-driven energy purchasing in smart grid cyber-physical system. *IEEE Transactions on Systems, Man, and Cybernetics: Systems, 43,* 773–784.

Timmerman, M. (2007). Embedded systems: Definitions, taxonomies, field.

Trapp, M., Schneider, D., & Liggesmeyer, P. (2013). A safety roadmap to cyber-physical systems. In *Perspectives on the future of software engineering.* Springer.

Tripakis, S., & Sengupta, R. (2014). Automated intersections: A CPS grand challenge. In *National Workshop on Transportation Cyber-Physical Systems* (p. 1), Arlington, Virginia.

Tsang, R. (2010). Cyberthreats, vulnerabilities and attacks on SCADA networks. In *Working Paper.* Berkeley: University of California. http://gspp.berkeley.edu/iths/Tsang_SCADA%20Attacks.pdf. As of December 28, 2011.

Twente, U. O. (2016). *Embedded systems* [Online]. The Netherlands: University of Twente. Available: https://www.utwente.nl/emsys/general/general/. Accessed January 25, 2016.

Union, E. (2016). *Cyber-physical systems* [Online]. Communications Networks, Content and Technology, European Commission Directorate General. Available: https://ec.europa.eu/dgs/connect/en/content/cyber-physical-systems-european-ri-strategy. Accessed January 19, 2016.

Wan, J., Chen, M., Xia, F., Di, L., & Zhou, K. (2013). From machine-to-machine communications towards cyber-physical systems. *Computer Science and Information Systems, 10,* 1105–1128.

Wan, J., Yan, H., Suo, H., & Li, F. (2011). Advances in cyber-physical systems research. *KSII Transactions on Internet and Information Systems (TIIS), 5,* 1891–1908.

Wang, J., Abid, H., Lee, S., Shu, L., & Xia, F. (2011). A secured health care application architecture for cyber-physical systems. *Control Engineering and Applied Informatics, 13,* 101–108.

Wang, W., & Lu, Z. (2013). Cyber security in the Smart Grid: Survey and challenges. *Computer Networks, 57,* 1344–1371.

Wasicek, A., Derler, P., & Lee, E. A. (2014). Aspect-oriented modeling of attacks in automotive Cyber-Physical Systems. In *Design Automation Conference (DAC), 2014 51st ACM/EDAC/ IEEE* (pp. 1–6). IEEE.

Weiser, M. (1991). The computer for the 21st century. *Scientific American, 265,* 94–104.

Wu, G., Lu, D., Xia, F., & Yao, L. (2011). A fault-tolerant emergency-aware access control scheme for cyber-physical systems. *arXiv preprint* arXiv:1201.0205.

Xia, F., Vinel, A., Gao, R., Wang, L., & Qiu, T. (2011). Evaluating IEEE 802.15. 4 for cyber-physical systems. *EURASIP Journal on Wireless Communications and Networking, 2011,* 596397.

Yan, Z., Zhang, P., & Vasilakos, A. V. (2014). A survey on trust management for Internet of Things. *Journal of Networks and Computer Applications, 42,* 120–134.

Ye, H. (2015). Security protection technology of cyber-physical systems. *International Journal of Security and Its Applications, 9,* 159–168.

Zhang, L., He, J., & Yu, W. (2012). Challenges and solutions of cyber-physical systems. *Information Science and Industrial Applications, 55.*

Zheng, X., Julien, C., Kim, M., & Khurshid, S. (2014). On the state of the art in verification and validation in cyber physical systems.

Zhou, K., Liu, B., Ye, C., & Liang, L. (2014). Design support tools of cyber-physical systems. In *Cloud computing.* Springer.

Zhu, B., Joseph, A., & Sastry, S. (2011a). A taxonomy of cyber attacks on SCADA systems. In *International Conference on Internet of Things (iThings/CPSCom), and 4th International Conference on Cyber, Physical and Social Computing* (pp. 380–388). IEEE.

Zhu, Q., Rieger, C., & Başar, T. (2011b). A hierarchical security architecture for cyber-physical systems. In *4th International Symposium on Resilient Control Systems (ISRCS)* (pp. 15–20). IEEE.

Zimmer, C., Bhat, B., Mueller, F., & Mohan, S. (2010). Time-based intrusion detection in cyber-physical systems. In *Proceedings of the 1st ACM/IEEE International Conference on Cyber-Physical Systems* (pp. 109–118). ACM.

Zimmer, C., Bhat, B., Mueller, F., & Mohan, S. (2015). Intrusion detection for CPS real-time controllers. In *Cyber physical systems approach to smart electric power grid.* Springer.

Zometa, P., Kogel, M., Faulwasser, T., & Findeisen, R. (2012). Implementation aspects of model predictive control for embedded systems. In *American Control Conference (ACC)* (pp. 1205–1210). IEEE.

Chapter 7
Distributed Control Systems Security for CPS

Distributed control systems (DCSs) are one of the founding technologies of the cyber-physical systems (CPSs), which are implemented in industries and grids. The DCSs are studies from the aspect of design, architecture, modeling, framework, management, security, and risk. From the findings, it was identified that the security of these systems is the most vital aspect among the modern issues. To address the security of DCSs, it is important to understand the bridging features between DCSs and the CPSs in order to protect them from cyberattacks against known and unknown vulnerabilities.

7.1 Introduction

The distributed control systems form the backbone of the modern industries (Stouffer et al. 2011). DCSs can be found in automobiles, airplanes, large industries, water management, power plant, refineries, flight management, healthcare systems, smart grids, etc. These systems consist of instruments that control the processes while taking care of the characteristics like accuracy, sensitivity, stability, reliability, speed, noise reduction, and bandwidth. To achieve the required goals, DCS implements physical sensors and actuators that are combined with sophisticated processors to form a CPS. As it is common in most other CPSs, the security issue is the most vital component of the DCSs (Stouffer et al. 2011). This is due to the implementation of DCS in critical infrastructures and other key industries that draw the attention of the adversaries and hackers. Although various countermeasures have been designed for generic information systems and CPS, they do not serve the DCS well and requires a distinctive approach toward security (Bologna et al. 2013). Also, the concept of DCS security is a new paradigm as compared to the information systems and thus it required extensive research and development (Weiss 2010).

© Springer International Publishing AG 2018
S. Ali et al., *Cyber Security for Cyber Physical Systems*, Studies in Computational
Intelligence 768, https://doi.org/10.1007/978-3-319-75880-0_7

7.2 Distributed Control Systems

Distributed control systems are addressed in various ways by different researchers. Some of the terms used for DCS are networked control systems (NCSs), industrial control systems (ICSs), and the networked cyber-physical systems (NCPSs). The NCPS is defined as a tight coupling of closed-loop control and actuation (Mangharam and Pajic 2013). The closed-loop property of NCPS signifies that it is equipped with feedback capability. For correct functioning, the NCPS must be reactive to the changes in the system and must maintain a situational awareness using the essential feedbacks (Stehr et al. 2010). Some of the applications of DCS are listed in Table 7.1.

The DCS is a combination of communication networks, computer science, computation, and control theory (Ge et al. 2015). DCS is also a process-oriented system that is limited in terms of its size and geographical distribution (Alcaraz and Zeadally 2015). The distinct characteristic of the DCS is that the overall system consists of large number of simpler subsystem/agents that are physical distributed

Table 7.1 Applications of distributed control systems

	Paper	Year	Country	Journal/conference	Description
1	Qian et al. (2015)	2015	China	International Journal of Hybrid Information Technology	Cyber-physical systems technology based on hybrid NC control system for automatic line
2	Bolognani et al. (2015)	2015	USA	IEEE Transactions On Automatic Control	A feedback strategy was proposed in smart power distribution grid for a distributed control law for optimal reactive flow
3	Zhong and Nof (2015)	2015	USA	Computers & Industrial Engineering	Smart water distribution network (WDN), systematic understanding of the collaborative response. Compared different parametric settings for performance measures (response time, maximum cascade, travel distance by responders, and preventability)
4	Giordano et al. (2014)	2014	Italy	International Conference on Internet and Distributed Computing Systems	Distributed and decentralized real-time approach methodology to control an urban drainage network
5	Loos et al. (2011)	2011	USA	International Symposium on Formal Methods	Proposed distributed car control system

and can interact with one another to coordinate the smaller tasks for achieving the desired collective objective (Ge et al. 2015). In each of the subsystems, the information is exchanged among its components (such as sensors and actuators). One of the major applications of DCS is in the power and energy sector. The DCS is implemented from the generation to the distribution of the electricity. Other applications of DCS range from critical infrastructures to large-scale industrial applications like water, transportation, health care, defense, and finance (Sandberg et al. 2015). Some of the main challenges to DCS in the current times are the communication, control, computation, and security. There is an intense tradeoff among some key issues like flexibility, efficiency, environmental, sustainability, distributed QoS, security, and the cost for the services (Ilic et al. 2010). However, security can be considered as the most vital issue in the DCS (Sandberg et al. 2015; Knapp and Langill 2014). Breach of security can create disruptions in the control, communication, and/or computations of the systems. Also, security breaches can lead to disasters in terms of monetary, information, and property damages. The modern DCSs are being connected to the internet for the advantages of remote access, efficiency, and ubiquity. This has led to the annexation of the DCS into the realm of CPS. Thus, this research concentrates on the DCS from a perspective of CPSs.

The distributed control system contains various components, layers, and specifications based on the requirements and implementation. The layers in a typical DCS are shown in Fig. 7.1. The operational layer consists of the control room and the central computer, monitoring the activity in the system. The master control is equipped with various servers to store the data and control action logs. Dedicated servers may be used for Web, video, simulation, and execution. The control level is maintained by supervisory computers and programmable logic controllers that provide the control signals to the concerned devices. A distributed network of microcontrollers that are physically connected to the actuators and other moving parts forms the direct control level. The plants and all of its working components constitute the field level that performs the tasks based on the control signals provided by the higher level to achieve the desired goal.

7.3   Design and Architecture

Designing of the distributed control system is the first step toward building and implementing a system that can be controlled in a distributed manner. The architecture of a typical DCS for electrical grids consists of generation unit with its sensors and actuators, the transmission unit with its gateways, the distribution unit with its gateways, and the customer section with its appliances (Alcaraz et al. 2016). All these units are managed using the super-nodes. The current literature shows that the modern focus is on forming a hybrid design and architecture that can integrate the synchronous with the asynchronous, physical with cyber, and higher level with lower. One of the applications of distributed CPS is in the aeronautics systems.

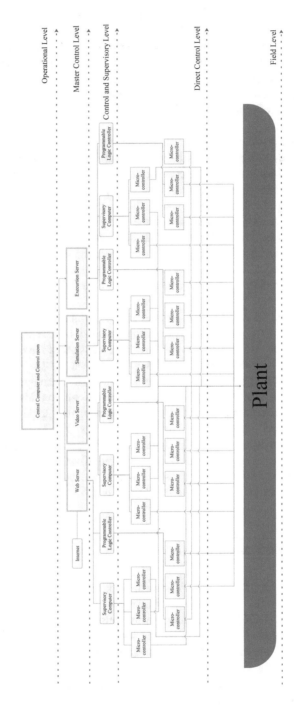

Fig. 7.1 Layers and components in a distributed control system

These systems consist of a collection of digital sensors and actuators that communicate to each other asynchronously and interact with the environment. In case of a pilot turning the airplane to a desired direction, the DCSs are implemented. The multirate physically asynchronous but logically synchronous (PALS) technique is a methodology that reduces the design and verification overheads by making the process simpler (Bae et al. 2015). PALS methodology is suited for systems that have various constraints and are physical asynchronous in their actions but the logical design of those systems is synchronous so that every component works together toward an objective.

DCS is also implemented in robotics since robots trying to resemble humans consist of a distributed number of sensors and actuators. The multilevel and hybrid architecture is used for communication and control of intelligent mobile robots (Posadas et al. 2008). The architecture is composed of three distributed parts, namely deliberative, reactive, and communication part. A communication framework known as SC agent implements the hybrid multilevel and distributed architecture. The distributed agents run on processors and workstations to control the mobile robots. The real-life implementation on the YAIR robot demonstrated effectiveness of the distributed agent in real-time executions (Posadas et al. 2008).

7.4 Modeling and Framework for DCS and CPS

Modeling helps to understand the design and the architecture while the framework leads to the practical functioning of the model. There are a number of models used in representing DCS and CPS. The cyber-based dynamic model is a model that uses mathematical modeling and depends on the cyber technologies connected to the physical system (Ilic et al. 2010). The model ensures the observability of the components in the systems by facilitating co-operative interaction among the distributed observers (Khan et al. 2008). The co-operation is achieved using the distributed iterative-collapse inversion algorithm (Khan and Moura 2008). The distributed sensor localization is based on the convergence algorithm (Khan et al. 2009). The model consisted of local sub-objectives that are built using distributed sensors and actuators. A three-step technique helps attain stability of performance in the distributed system. The model enables interactive protocols between the controllers and operators and enables distributed decision-making. The complex power system operations are assisted using data mining. The model is an initiative toward managing the tradeoffs between the issues like flexibility, efficiency, sustainability, distributed quality of service, and security. Table 7.2 describes the various researches being carried on the modeling and design of DCSs.

Distributed control systems must be reactive, maintain situational awareness, and must have essential feedbacks (Stehr et al. 2010). For addressing the challenges, a declarative approach could be implemented to avoid the low-level programming and its error-prone and time-consuming nature. Using the declarative method, the

Table 7.2 Modeling and design of distributed control systems

	Paper	Year	Country	Journal/conference	Description
1	Bae et al. (2015)	2015	USA	Science of Computer Programming	Multirate PALS, real-time maude, model checking, hybrid systems
2	Zhu and Basar (2015)	2015	USA	IEEE Control Systems Magazine	Hybrid game theoretic framework was presented where stochastic switching and deterministic uncertainties are represented by the known range of disturbances through modeling the occurrence of unanticipated events
3	Bradley and Atkins (2015)	2015	USA	IEEE Transactions on Robotics	New ideas and innovative approaches to develop novel models and ideas to couple interacting cyber and physical strategies
4	Li et al. (2015)	2015	USA	International Conference on Cyber-Physical Systems	Slot stealing and event-based communication protocols were proposed for efficient real-time emergency communication
5	Kim and Kumar (2012)	2012	USA	Proceedings of the IEEE	CPS research presenting historical overview on early generations of control systems technologies and potentials for CPS in many application domains
6	Morris et al. (2011)	2011	USA	Cyber Security and Information Intelligence Research Workshop	Development of lab equipped with commercial software and hardware devices provided by multiple vendors and perform unique test bed for industrial control system, discovery of cyber vulnerabilities, and solutions research

collected information, control activities, and decision-making are all converted to logical problems. The logic captures the interactions with the physical world.

A reasoning that is a distributed process in space and time is a paradigm shift from the traditional method. The logical problems are events of the system and are based on distributed knowledge which consists of equally treated facts and goals. The knowledge is opportunistically disseminated every time the robots are connected with the sensors. A logical framework is created by combining network distributed reasoning by partially ordered knowledge sharing and asynchronous control. The framework was then implemented on a simulation prototype in the setting of self-organizing mobile robot teams. The situation taken into consideration was the one of emergency where the robots are highly dynamic and frequently failure prone which needs to be compensated by real-world actions. The system continuously carries out a distributed reasoning to compute the local solution and make the movement adjustments without interruptions. Although the proposed framework and its simulation showed good performance, it needs to be tested on real-world implications and compared to other frameworks in the field.

The other concerns in the modeling are the communication infrastructure and data traffic in the DCS. Communication infrastructure is considered as one of the prominent parts of the CPS since it carries the information from sensors to the controllers. The data traffic is used for the controlling of the system dynamics. The scheduling takes care of the medium access layer which is concerned with the selection of communication channels and it plays an important role in the efficiency of the CPS. The current literature demonstrates a scheduling framework based on hybrid systems (Li et al. 2014). The hybrid system is a system that consists of both continuous and discrete systems and can be switched based on the system state and control. The framework takes care of the cyber and physical dynamics of the CPS and selects the channel accordingly. The framework was applied in the voltage control of distributed generations. Distributed generation is a system that contains multiple sources of electricity from generators using renewable and nonrenewable resources. The simulations showed to attain satisfactory performance results. The framework needs to be implemented in other domains of CPS to prove its feasibility.

7.5 Management of DCS

The DCS or networked controlled system is a tight coupling of the control and actuation of physical processes in a closed loop (Mangharam and Pajic 2013). Its management requires efficient and smart use of resources and components based on the parameters being sensed. Effective management facilities sustaining desired stability and performance of the DCS. The management must be equipped with robust, reliable, and optimal control to provide the best performance. The current researches on the management and control aspect of the DCS are listed in Table 7.3.

One of the tools for management of DCS is the embedded virtual machines program (EVM) (Mangharam and Pajic 2013). It is a program in which task, control, and timing properties are maintained in the physical nodes of the DCS. It is capable of finding the best set of physical controllers during runtime to keep the system stable. Although the EVM provides the optimal configuration for the routing and task assignment, it is not capable of proving the guarantee for it. Distributed control over wireless network is another technique of DCS management. The wireless control network is a tool where the entire network is treated as a controller rather than each individual node in the network (Pajic et al. 2011). The control computations are done completely within the network using a truly distributed control scheme. The network thus behaves as a linear dynamical system. This system is more robust and stable in times of node failure. The tool also addresses the security issue by providing an intrusion detection system within its architecture. The tool lacked application in heterogeneous nodes with variable capabilities.

Table 7.3 Management and control in distributed control systems

	Paper	Year	Country	Journal/conference	Description
1	Alcaraz et al. (2016)	2016	Spain	Journal of Network and Computer Applications	A policy enforcement system based on the approach where degree of observations of a context- and role-based access control model defined by IEC 62351-8 Standard
2	Mocci et al. (2015)	2015	Italy	Electric Power Systems Research	Control of loads based multi-agent system (MAS) for DSI implementation
3	Zhang et al. (2015)	2015	USA	International Conference on Cyber-Physical Systems	Controller verification scheme for detecting unstable learning behaviors caused by control algorithm/software fault and unanticipated physical faults is developed for a class of first-order nonlinear adaptive control systems'
4	Zhu et al. (2013)	2013	USA	Lecture Notes in Control and Information Sciences	Resilient control design for multi-agent cyber-physical systems. General systems framework interactions for cyber and physical components within CPS and their dependencies among multiple CPSs
5	Mangharam and Pajic (2013)	2013		Journal of the Indian Institute of Science	Challenges identification of time-critical closed-loop control problems by the use wireless networks
6	Stehr et al. (2010)	2010	USA	International Conference on Ubiquitous Intelligence and Computing	Declarative control of NCPS through distributed computational and logical foundation
7	Li et al. (2014)	2014		IEEE Systems Journal	Use of scheduling algorithms for data traffic for controlling system dynamics in CPS

(continued)

Table 7.3 (continued)

	Paper	Year	Country	Journal/conference	Description
8	Colombo et al. (2014)	2014	Switzerland	Industrial Cloud-Based Cyber-Physical Systems	Monitoring and controlling industrial applications (lower levels within the plant hierarchy) include PLC, SCADA, and DCS systems. Industrial processes and critical infrastructures and their complex functionalities depend upon SCADA and DCS systems
9	Harrison et al. (2014)	2014	Switzerland	Industrial Cloud-Based Cyber-Physical Systems	Any level of ISA-95 can be integrated using one or more specified tools to provide coherent SOA-based SCADA/DCS solution
10	Stouffer et al. (2011)	2011	USA	Guide to Industrial Control Systems Security (NIST Special Publication 800-82)	Configuration guide is presented for SCADA, DCS, and other control systems like program logic controllers (PLC)
11	Zhang and Chow (2012)	2012	USA	IEEE Transactions on Power Systems	To control to minimize the total cost of operations in the power systems, a use of consensus algorithm embedded in generation units as an effective means of distributed control is explored
12	Posadas et al. (2008)	2008	Spain	Engineering Applications of Artificial Intelligence	Modular and general hybrid architecture was developed to control different types of systems particularly mobile robot control systems

The control operations are simple ones, effectivity with complex control operations was not demonstrated.

In case of the smart grids, the major management issue is the management the economic dispatch problem (EDP). Economic dispatch is the process of determining the most optimal output to be generated by the electrical generators to meet the demand with the lowest cost in the given constraints of transmission and operation (Zhang and Chow 2012). The distributed algorithm to solve the EDP is used to replace the traditional central controller. The higher stability of DCSs forms an advantage over the centralized control system. The basic problem in DCS is the requirement for all the nodes to reach a consensus over a given issue. To address this issue, the incremental cost convergence algorithm is embedded with the generation unit of the DCS (Zhang and Chow 2012). The algorithm also provides a procedure to appoint the leader in the group of nodes using the node centrality measurement technique. Each cluster is also equipped with the consensus manager that provides the interactions between the clusters. Using a three-bus micro-grid, different network topologies like star connection, random connection, and serial connection were simulated and studied. The simulations demonstrated effectiveness and robustness as compared to the centralized control. Although the simulations proved to show better performance, the results were not verified with real-life implementations.

The ever increasing climate change and the urbanization led to the drainage systems being overwhelmed. This may eventually lead to sewer flooding and widespread damages to human lives, properties, and environment. To address this issue, a decentralized and distributed agent-based approach was developed to manage the drainage system (Giordano et al. 2014). The approach is based on the gossip-based algorithm (Jelasity et al. 2005) and the PID controlling technique to realize the real-time control. A distributed system of sensors and actuators is set up for real-time control. Each node is able to communicate only with its neighboring peers. The gossip-based algorithm continuously monitors and balances the water level. Each gate is controlled locally using the PID controllers. The simulation experiments with the proposed approach showed that it was able to prevent or delay flooding in the given scenarios. This model was not validated in the real-world drainage networks.

7.6 Security and Risk

Security can be defined as to protect an asset from its vulnerabilities and threats. The assets could be human, data, systems, organization, country, etc. Information security is to protect information from unauthorized access, modifications, information leakage, perusal, disruption, and destruction (Jagadamba et al. 2014; Felderer et al. 2014; Ansari and Janghel 2013; House 2014). The generic security issues are integrity, confidentiality, and availability. The CPS and DCS have added issues of authenticity and validation (Kriaa et al. 2015).

7.6.1   Security Issues

7.6.1.1   Integrity

Integrity refers to the trust of truthfulness on the data or resources in the system (Cardenas et al. 2008). The integrity is the guarantee that the system will perform the tasks intended by the designer or user. The data and the information present within the system are vital for the sensing and computing and the decision-making processes. Thus, the integrity of the data is very critical in any system. An attacker having an access to the data can easily manipulate the data based on his/her requirements and thus achieving a physical action which is desired by the attacker rather than the expected action. In the context of DCS, to ensure smooth functioning, the integrity must be maintained at the compound, device, and bus level (Rauter 2016).

7.6.1.2   Confidentiality

It refers to the ability of keeping all the information secret from outsiders and attackers (Cardenas et al. 2008). Sometimes, though the attacker may not be able to manipulate the data, it may be possible to just sense the data being sent and received. This data may be confidential and secret, and this could enable the attacker to take necessary actions against the information obtained for its own benefit. In case of DCS, the data and the control traffic is required to maintain confidentiality (Innovations 2014).

7.6.1.3   Availability

The service by a system is expected to be delivered in every instance of its need. Especially, it is very important in critical systems like smart grid or nuclear reactors where unavailability of even a fraction of a second can be fatal. An attacker can exploit this critical aspect to create a disaster by keeping the system occupied in some worthless processes while the critical process is being delayed (Pappas et al. 2008; Solomon and Chapple 2009). The DCS must be equipped with end-to-end security for the communication protocol to ensure availability (Hieb et al. 2007).

7.6.1.4   Authenticity and Validation

DCS and CPS always involve communication between various entities and it is very crucial to ensure that the entities are who they claim they are. This necessitates the authenticity that the data and transactions and validation of the received information (Kriaa et al. 2015). The operating system of the DCS must be robust to ensure the authenticity and validity of the data, control, and nodes (Sinopoli et al. 2003).

7.6.1.5 Security Attacks

Some of the major network-based attacks have been discussed below.

Denial of Service

It is one of the general forms of cyberattack against smart grid. It is called denial of service because it denies normal services being accessed by legitimate users. Denial-of-service attack is developed through massive resources exhaustion which floods the communication on a particular network for being inaccessible or attack server providing such services with huge volumes of traffic or spurious workloads (Habash et al. 2013; Govindarasu et al. 2012).

Eavesdropping

Eavesdropping can be achieved through monitoring and obtaining sensitive information about an adversary ended up with privacy breaches by stealing power usages, disclosure of the controlling structure of smart grids. Commonly, eavesdropping techniques are being used for information gathering for further attacks. In smart grid environment, an attacker can gather and examine network traffic to deduce information from communication patterns and may make communication inaccessible for traffic analysis attacks.

Man-in-the-Middle

This kind of attack involved third party who eavesdrop the communication between two legitimate entities where attacker makes independent connection with the victims and relays messages between them to ensure the reliability of the communication. In reality, the whole communication is controlled by the attacker.

Time Synchronization Attack

This kind of attack targets the timing information in the smart grid infrastructure (SGI) to effect transmission line fault detection, event localization and voltage stability monitoring, and three applications of phasor measurement units (Aloul et al. 2012).

Routing Attacks

Routing attacks involved the cyberattacks on routing infrastructure of the network.

Malware

Malicious software exploits the vulnerabilities in control systems through its system software, PLCs, or protocols. This kind of attack scans the network to find vulnerabilities within the victim machines and replicates the malware payload for self-propagation.

Network-Based Intrusion

This kind of attacks exploiting network through poorly designed or configured firewalls for both misconfigured inbound and faulty outbound rules to allow adversary to inject malicious payload into the control system for desirable objectives.

7.6.2 Security Measures

The modern DCSs use the corporate networks and the internet for its operation. This has led to an increased exposure to the cyber-threats and vulnerabilities (Knowles et al. 2015). This is proved by the fact that in 2015, the ICS-CERT reported 24 times more vulnerabilities than in 2010 times and it was double when compared to 2014 (Felker and Edwards 2015). There was 20% increase in cyber incidents in 2015 from that in 2014 (Felker and Edwards 2015). The security issues and security attacks on the DCS directed to the formulation of the security measures and countermeasures. The security measures are expected to defend and protect the DCS from security breach and violations. Realizing the importance of the DCS and CPS, various industries, government, standardization organization, and researchers have formulated numerous security measures, standards, guidelines, and best practices. Majority of the standards were found to be US based (Knowles et al. 2015). In a DCS, the cyber components and the physical components are interconnected and due to this, their security becomes interdependent. The security also primarily depends on the human actions and decisions-making over the system from insiders and outsiders. The NCSs are subjected to various attacks like stealth, replay, covert, and false data injection. There is a need to develop tools to analyze and synthesize the combination of control theory, game theory, and network optimizations for DCS (Sandberg et al. 2015). Some of the known researches on DCS security are listed in Table 7.4.

One of the most vulnerable forms of DCS is the SCADA (Teixeira et al. 2012). This is mainly due to the unprotected channels and the feedback loops present in the SCADA networks (Teixeira et al. 2012). The model of the attacks on such system can be viewed in a three-dimensional space with the axes being system knowledge, disclosure resources, and disruption resources (Teixeira et al. 2012). The system knowledge is the amount of knowledge the attacker has about the control system's core components. The disclosure resource is the sequences of data collected by the attacker from the calculated control actions. The disruption resources give the attack vector that can be used to affect the components of the system. This framework is found to be suitable for the replay, zero dynamics, and bias injection attack scenarios. The model was incapable of analyzing other attacks like Sybil attacks, eavesdropping, denial of service, etc. The cross-layer system model was developed to study the multi-agent environment and decentralized nature of the DCS (Zhu et al. 2013). This model facilitates the study and analysis of the performance and coupling in CPS. A mischievous agent tries to mislead the entire system to perform undesirable actions used the model to study the attack. The feedback Nash equilibrium technique based on game theory is used as a solution for this kind of attack scenario (Zhu et al. 2013). The mechanism is used to compute the distributed control strategies for each unit of the system. The framework needs to be practically implemented to validate the expected performance improvements.

An unauthorized and unauthenticated connection within a DCS is another major security issue. In smart grids, a critical infrastructure, connections may arise at any

Table 7.4 Research papers on distributed control systems security

	Paper	Year	Country	Journal/conference	Description
1	Alcaraz and Zeadally (2015)	2015	Spain	International Journal of Critical Infrastructure Protection	Technology trends and security issues, trust management, and privacy. Focuses on security of industrial control systems for the integration of new technologies with legacy systems
3	Teixeira et al. (2012)	2012	Swedish	International Conference on High Confidence Networked Systems	Analysis of security measures for network control systems
2	Cárdenas et al. (2008)	2008	USA	28th International Conference on Distributed Computing Systems Workshops (IEEE)	Discussed secure control systems, robust network control system, fault tolerance control, and make four claims for what missing in cps secure control system
4	Boyer and Mcqueen (2007)	2008	USA	International Workshop on Critical Information Infrastructures Security	Proposed security metrics for distributed control system used in chemical processing plant
5	Ralston et al. (2007)	2007	USA	ISA Transactions	Discussed issues related with cybersecurity of SCADA and DCS networks used to control critical infrastructures

time, from anywhere, and in anyway. It is vital that a policy is enforced that protects the smart grid from suspicious connections. Smart grid technology requires high amount of interoperability in their operations (Miller 2010). Interoperability will allow the smart grid system to share and use the available information effectively and efficiently to take the correct design and perform the expected task.

To enable the interoperability, a policy enforcement system is used for transparent control operations in a safe, secure, and reliable architecture (Alcaraz et al. 2016). The policy uses context-based approach, graph theory, and role-based access control. The authentication is carried out in stage 1, authorization in stage 2, and interoperability in stage 3 of the policy architecture. The approach was tested using simulations for various scenarios and proved to be effective. The approach did not take into consideration of the faults that may arise during the runtime of the control systems.

The safety is one of the major concerns in the field of transportation. Various researches are concentrated on the effectiveness of the safety measures in the cars.

It has been found that the effectiveness is the most when the control of the vehicle is carried out in a distributed manner using distributed sensing, communication, and decision-making. A model is thus implemented for distributed car control where adaptive cruise control is used to control every car in the system (Loos et al. 2011). The model organizes the cars in the vicinity into a collection of hierarchical modular pieces of two cars on a single lane in one packet. Using the distributed control model, the car is provided safety at local and global levels. The model was formally proved in various complex settings to ensure its safety objectives by providing guaranteed collision freedom from other cars. Some of the limitations are that the model was not verified in the real-world implementation. The model also does not take care of the issues of time synchronization, sensor data inaccuracy, and curvature of the lanes.

7.7 Risk

Oracle Corporation has defined risk as the possibility of loss, damage, or any other undesirable event (Corporation 2008). PricewaterhouseCoopers (PwC) defines it as "the possibility that an event will occur and adversely affect the achievement of objectives" (PricewaterhouseCoopers 2008). Thus, risk is an undesirable but unavoidable possibility of occurring of an event and this requires an assessment of the risks involved in any given system. As the dependence of critical infrastructures and industrial automation on cyber-physical control systems is growing, many unforeseen security threats to DCS are unfolding. A DCS may be subjected to many vulnerabilities, threats, and security issues which may lead to catastrophic events in the society. This requires an efficient and effective risk management system to maintain the minimum possible risk in the system. As presented by Cho et al (2011), risk management is constituted by three parts, namely risk assessment, risk mitigation, and risk control. It is critical that engineers, managers, and operators to understand the issues in these three parts and know how to locate the information they need.

The security metrics of control systems consists of seven risk management ideals, namely the security group, attack group, access, vulnerabilities, damage potential, detection, and recovery (Boyer and McQueen 2007). Each of these ideals consists of principles and best practices that could assist in risk measurement and risk avoidance. A DCS system based on TCP/IP and consisting of 30 distributed controllers was successfully tested with the above principles and metrics by Boyer and McQueen using the mathematical modeling (Boyer and McQueen 2007). The paper could not provide the effectiveness of the model in terms of component test count, attack surface, and detection performance. It also failed to present a technique to provide a quantitative measure of the security risks in DCS. Based on another study, it was concluded that the risk management process consisted of six steps, namely context identification, risk assessment, risk estimation, risk evaluation, risk treatment, and risk acceptance (Knowles et al. 2015). A six-step framework was also developed by a different group to reduce the security risk in DCS (Ralston et al. 2007). It starts with the construction of the vulnerability tree and conducting an effect analysis and threat impact on them. Based on them, threat-impact index and vulnerability index are computed and added into the tree to complete the tree. The current literature shows that there is a lack of practical assessment methodologies. The future research areas in the domain of security management could be the component security, real-time risk assessment, system-wide security assessment, security control efficacy, and the interdependence modeling. It is also required to find a method to qualify the risk of each component in the system.

Some of the risk assessment techniques being implemented on CPS and capable to be used on DCS are hierarchical holographic modeling (Haimes 2015), risk filtering raking and management (Haimes et al. 2002), and inoperability input–output model (Liu and Xu 2013). Some of the tools that assist in the risk assessment are the RiskWatch, operationally critical threat asset and vulnerability evaluation, Proteus, and the CORAS.

7.8   Conclusion

Distributed control system is closely related to the CPSs. The applications of DCS range from large-scale industrial applications to critical infrastructures like water, transportations, electricity, health care, defense, and finance. The issue in this domain is the intersection of the control systems and the computer security. The critical structures governed DCS are subjected to both cyber and physical attacks. This requires robust security protection and systematic risk assessment techniques to assure the safety and availability of the critical infrastructures. Its importance is proved by the initiatives taken by the Department of Homeland Security, Government of the United States. The governments of both USA (2002) and UK (2008) have created guidelines to deal with the DCSs.

This chapter makes an attempt to study the DCS with respect to the CPSs. The issues concerning designing, modeling, architecture, and management have been studied and discussed. The security issue forms the most vital aspect of the DCS and has been given a great importance in this domain. The current state of the DCS demands a security and risk framework that can address the physical as well as the distributed aspect of the system. Since the system functions in a distributed manner rather than a centralized one, it is understood that each node in the network is given certain communication and decision-making capabilities. Thus, for security purposes, it is essential to maintain a trust and reputation among the nodes. The concept of trust and reputation have not yet been implemented or studied in the domain of DCS. It is required that the future research focuses on a combination of trust, reputation, and risk management framework that is expected to improve the security and performance of the DCS.

References

Alcaraz, C., Lopez, J., & Wolthusen, S. (2016). Policy enforcement system for secure interoperable control in distributed smart grid systems. *Journal of Network and Computer Applications, 59,* 301–314.

Alcaraz, C., & Zeadally, S. (2015). Critical infrastructure protection: Requirements and challenges for the 21st century. *International Journal of Critical Infrastructure Protection, 8,* 53–66.

Aloul, F., Al-Ali, A., Al-Dalky, R., Al-Mardini, M., & El-Hajj, W. (2012). Smart grid security: Threats, vulnerabilities and solutions. *International Journal of Smart Grid and Clean Energy, 1,* 1–6.

Ansari, S., & Janghel, R. R. (2013). A dynamic approach to generate behavior patterns of virus and worms for intrusion detection system. *International Journal of Advanced Research in Computer Science, 4.*

Bae, K., Krisiloff, J., Meseguer, J., & Ölveczky, P. C. (2015). Designing and verifying distributed cyber-physical systems using Multirate PALS: An airplane turning control system case study. *Science of Computer Programming, 103,* 13–50.

Bologna, S., Fasani, M. A., & Martellini, M. (2013). The importance of securing industrial control systems of critical infrastructures. *General Secretariat.* Como, Italy: Landau Network. Retrieved January, 14, 2014.

Bolognani, S., Carli, R., Cavraro, G., & Zampieri, S. (2015). Distributed reactive power feedback control for voltage regulation and loss minimization. *IEEE Transactions on Automatic Control, 60,* 966–981.

Boyer, W., & Mcqueen, M. (2007). Ideal based cyber security technical metrics for control systems. In *International Workshop on Critical Information Infrastructures Security* (pp. 246–260). Springer.

Bradley, J. M., & Atkins, E. M. (2015). Coupled cyber-physical system modeling and coregulation of a cubesat. *IEEE Transactions on Robotics, 31,* 443–456.

Cardenas, A. A., Amin, S., & Sastry, S. (2008). Secure control: Towards survivable cyber-physical systems. *System, 1,* a3.

Cárdenas, A. A., Amin, S., & Sastry, S. (2008). Research challenges for the security of control systems. In *HotSec.*

Cho, J.-H., Swami, A., & Chen, I.-R. (2011). A survey on trust management for mobile ad hoc networks. *Communications Surveys & Tutorials, IEEE, 13,* 562–583.

Colombo, A. W., Karnouskos, S., & Bangemann, T. (2014). Towards the next generation of industrial cyber-physical systems. In *Industrial cloud-based cyber-physical systems.* Springer.

Corporation, O. (2008). Risk Analysis Overview. http://www.oracle.com/us/products/middleware/bus-int/crystalball/risk-analysis-overview-404902.pdf, Date accessed: 6 /10/ 2015

Felderer, M., Katt, B., Kalb, P., Jürjens, J., Ochoa, M., Paci, F., et al. (2014). Evolution of security engineering artifacts: A state of the art survey. *International Journal of Secure Software Engineering (IJSSE), 5,* 48–98.

Felker, J., & Edwards, M. (2015). *NCCIC/ICS-CERT year in review.* FY 2015.

Ge, X., Yang, F., & Han, Q.-L. (2015). Distributed networked control systems: A brief overview. *Information Sciences.*

Giordano, A., Spezzano, G., Vinci, A., Garofalo, G., & Piro, P. (2014). A cyber-physical system for distributed real-time control of urban drainage networks in smart cities. In *International Conference on Internet and Distributed Computing Systems* (pp. 87–98). Springer.

Govindarasu, M., Hann, A., & Sauer, P. (2012). Cyber-physical systems security for smart grid. In *The future grid to enable sustainable energy systems.* PSERC Publication.

Habash, R. W., Groza, V., & Burr, K. (2013). Risk management framework for the power grid cyber-physical security. *British Journal of Applied Science & Technology, 3,* 1070.

Haimes, Y. Y. (2015). *Risk modeling, assessment, and management.* Wiley.

Haimes, Y. Y., Kaplan, S., & Lambert, J. H. (2002). Risk filtering, ranking, and management framework using hierarchical holographic modeling. *Risk Analysis, 22,* 383–397.

Harrison, R., McLeod, C. S., Tavola, G., Taisch, M., Colombo, A. W., Karnouskos, S., et al. (2014). Next generation of engineering methods and tools for SOA-based large-scale and distributed process applications. In *Industrial cloud-based cyber-physical systems.* Springer.

Hieb, J., Graham, J., & Patel, S. (2007). Security enhancements for distributed control systems. In *International Conference on Critical Infrastructure Protection* (pp. 133–146). Springer.

House, T. W. (2014). *Co-ordination of federal information security policy* [Online]. The United States Government. Available: https://www.whitehouse.gov/sites/default/files/omb/legislative/letters/coordination-of-federal-information-security-policy.pdf. Accessed July 15, 2016.

Ilic, M. D., Xie, L., Khan, U. A., & Moura, J. M. (2010). Modeling of future cyber–physical energy systems for distributed sensing and control. *IEEE Transactions on Systems, Man, and Cybernetics-Part A: Systems and Humans, 40,* 825–838.

Innovations, R.-T. (2014). *Four keys to securing distributed control systems.* California, US: Real-Time Innovations.

Jagadamba, G., Sharmila, S., & Gouda, T. (2014). A secured authentication system using an effective keystroke dynamics. In *Emerging research in electronics, computer science and technology.* Springer.

Jelasity, M., Montresor, A., & Babaoglu, O. (2005). Gossip-based aggregation in large dynamic networks. *ACM Transactions on Computer Systems (TOCS), 23,* 219–252.

Khan, U. A., Ili, M. D., & Moura, J. M. (2008). Cooperation for aggregating complex electric power networks to ensure system observability. In *First International Conference on Infrastructure Systems and Services: Building Networks for a Brighter Future (INFRA)* (pp. 1–6). IEEE.

Khan, U. A., Kar, S., & Moura, J. M. (2009). Distributed sensor localization in random environments using minimal number of anchor nodes. *IEEE Transactions on Signal Processing, 57,* 2000–2016.

Khan, U. A., & Moura, J. M. (2008). Distributed iterate-collapse inversion (DICI) algorithm for L-banded matrices. In *IEEE International Conference on Acoustics, Speech and Signal Processing* (pp. 2529–2532). IEEE.

Kim, K.-D., & Kumar, P. R. (2012). Cyber–physical systems: A perspective at the centennial. *Proceedings of the IEEE, 100,* 1287–1308.

Knapp, E. D., & Langill, J. T. (2014). *Industrial network security: Securing critical infrastructure networks for smart grid, SCADA, and other Industrial Control Systems.* Syngress.

Knowles, W., Prince, D., Hutchison, D., Disso, J. F. P., & Jones, K. (2015). A survey of cyber security management in industrial control systems. *International Journal of Critical Infrastructure Protection, 9,* 52–80.

Kriaa, S., Pietre-Cambacedes, L., Bouissou, M., & Halgand, Y. (2015). A survey of approaches combining safety and security for industrial control systems. *Reliability Engineering & System Safety, 139,* 156–178.

Li, B., Nie, L., Wu, C., Gonzalez, H., & Lu, C. (2015). Incorporating emergency alarms in reliable wireless process control. In *Proceedings of the ACM/IEEE Sixth International Conference on Cyber-Physical Systems* (pp. 218–227). ACM.

Li, H., Han, Z., Dimitrovski, A. D., & Zhang, Z. (2014). Data traffic scheduling for cyber physical systems with application in voltage control of distributed generations: A hybrid system framework. *IEEE Systems Journal, 8,* 542–552.

Liu, M., & Xu, W. (2013). The approach for critical infrastructure sectors classification using the inoperability input-output model (IIM). In *6th International Conference on Information Management, Innovation Management and Industrial Engineering* (pp. 7–10). IEEE.

Loos, S. M., Platzer, A., & Nistor, L. (2011) Adaptive cruise control: Hybrid, distributed, and now formally verified. In *International Symposium on Formal Methods* (pp. 42–56). Springer.

Mangharam, R., & Pajic, M. (2013). Distributed control for cyber-physical systems. *Journal of the Indian Institute of Science, 93,* 353–387.

Miller, C. (2010). *Interoperability and cyber security plan. NRECA CRN smart grid regional demonstration.* Arlington, Virginia, USA: Cigital Inc., Cornice Engineering Inc., Power Systems Engineering.

Mocci, S., Natale, N., Pilo, F., & Ruggeri, S. (2015). Demand side integration in LV smart grids with multi-agent control system. *Electric Power Systems Research, 125,* 23–33.

Morris, T., Vaughn, R., & Dandass, Y. S. (2011). A testbed for SCADA control system cybersecurity research and pedagogy. In *Proceedings of the Seventh Annual Workshop on Cyber Security and Information Intelligence Research* (pp. 27). ACM.

Pajic, M., Sundaram, S., Pappas, G. J., & Mangharam, R. (2011). The wireless control network: A new approach for control over networks. *IEEE Transactions on Automatic Control, 56,* 2305–2318.

Pappas, V., Athanasopoulos, E., Ioannidis, S., & Markatos, E. P. (2008). Compromising anonymity using packet spinning. In *International Conference on Information Security* (pp. 161–174). Springer.

Posadas, J. L., Poza, J. L., Simó, J. E., Benet, G., & Blanes, F. (2008). Agent-based distributed architecture for mobile robot control. *Engineering Applications of Artificial Intelligence, 21,* 805–823.

Pricewaterhousecoopers. (2008). *A practical guide to risk assessment.*

Qian, F., Xu, G., Zhang, L., & Dong, H. (2015). Design of hybrid NC control system for automatic line. *International Journal of Hybrid Information Technology, 8,* 185–192.

Ralston, P. A., Graham, J. H., & Hieb, J. L. (2007). Cyber security risk assessment for SCADA and DCS networks. *ISA Transactions, 46,* 583–594.

Rauter, T. (2016). Integrity of distributed control systems. In *Student Forum of the 46th Annual IEEE/IFIP International Conference on Dependable Systems and Networks.*

Sandberg, H., Amin, S., & Johansson, K. (2015). Cyberphysical security in networked control systems: An introduction to the issue. *Control Systems, IEEE, 35,* 20–23.

Sinopoli, B., Sharp, C., Schenato, L., Schaffert, S., & Sastry, S. S. (2003). Distributed control applications within sensor networks. *Proceedings of the IEEE, 91,* 1235–1246.

Solomon, M. G., & Chapple, M. (2009). *Information security illuminated.* Jones & Bartlett Publishers.

Stehr, M.-O., Kim, M., & Talcott, C. (2010). Toward distributed declarative control of networked cyber-physical systems. In *Ubiquitous intelligence and computing.* Springer.

Stouffer, K., Falco, J., & Scarfone, K. (2011). Guide to industrial control systems (ICS) security. *NIST Special Publication, 800,* 16–16.

Teixeira, A., Pérez, D., Sandberg, H., & Johansson, K. H. (2012). Attack models and scenarios for networked control systems. In *Proceedings of the 1st International Conference on High Confidence Networked Systems* (pp. 55–64). ACM.

UK. (2008). *Good practice guide—Process control and SCADA security* [Online]. London: Centre for the Protection of National Infrastructure. Available: http://www.cpni.gov.uk/documents/publications/2008/2008031-gpg_scada_security_good_practice.pdf?epslanguage=en-gb. Accessed May 11, 2016.

US. (2002). *21 steps to improve cyber security of SCADA networks* [Online]. Washington: US Department of Energy. Available: http://www.energy.gov/sites/prod/files/oeprod/DocumentsandMedia/21_Steps_-_SCADA.pdf. Accessed May 11, 2016.

Weiss, J. (2010). *Protecting industrial control systems from electronic threats.* Momentum Press.

Zhang, X., Clark, M., Rattan, K., & Muse, J. (2015) Controller verification in adaptive learning systems towards trusted autonomy. In *Proceedings of the ACM/IEEE Sixth International Conference on Cyber-Physical Systems* (pp. 31–40). ACM.

Zhang, Z., & Chow, M.-Y. (2012). Convergence analysis of the incremental cost consensus algorithm under different communication network topologies in a smart grid. *IEEE Transactions on Power Systems, 27,* 1761–1768.

Zhong, H., & Nof, S. Y. (2015). The dynamic lines of collaboration model: Collaborative disruption response in cyber–physical systems. *Computers & Industrial Engineering, 87,* 370–382.

Zhu, Q., & Basar, T. (2015). Game-theoretic methods for robustness, security, and resilience of cyberphysical control systems: games-in-games principle for optimal cross-layer resilient control systems. *IEEE Control Systems, 35,* 46–65.

Zhu, Q., Bushnell, L., & Basar, T. (2013) Resilient distributed control of multi-agent cyber-physical systems. In D. C. Tarraf (Ed.), *Lecture notes in control and information sciences* (pp. 301–316). The Johns Hopkins University, Springer.

Chapter 8
Standards for CPS

Cyber-physical system (CPS) has emerged as an enhancement to the existing generation of wireless sensor networks and embedded systems. The rapid growth in the applications of this platform calls for standardized guidelines that would ensure seamless operation and inter-compatibility between various components deployed in this environment. The components integrated within the CPS include sensing, computation, communication, and controlling units. Several institutes and organizations across the globe have attempted to address this concern by presenting a set of rules in the form of standards. In this chapter, standards are defined in general and why we need standards in the domain of CPSs is discussed in specific. Embedded systems in CPS will be presented as well as with cybersecurity standards, the available standards concerning CPS with other components will also be studied and analyzed.

8.1 Why Do We Need Standards in CPS?

Standards are defined as "technical specifications defining requirements for products, production processes, services or test-methods. These specifications are voluntary. They are developed by industry and market actors following some basic principles such as consensus, openness, transparency and non-discrimination. Standards ensure interoperability and safety, reduce costs and facilitate companies' integration in the value chain and trade" (CEN 2016a). Another definition for standard by European Telecommunication Standard Institute (ETSI 2016) is "a document, established by consensus and approved by a recognized body, that provides, for common and repeated use, rules, guidelines or characteristics for activities or their results, aimed at the achievement of the optimum degree of order in a given context".

A standard can also be defined as a published document that had been constructed by a consensus from the major scientists and then approved by a

© Springer International Publishing AG 2018
S. Ali et al., *Cyber Security for Cyber Physical Systems*, Studies in Computational Intelligence 768, https://doi.org/10.1007/978-3-319-75880-0_8

recognized body. Standards provide requirements, specifications, rules, guidelines, and procedures for a particular task, product, system, or service. They help in achieving the best extent of order, reliability, safety, consistency, and performance by providing a common language which defines quality and safety criteria (Australia 2015; IEC 2015, ISO, Commission).

Standards are developed to meet company, local, regional, or global application. It is also created for particular product or services specification. Standards can be applied on a voluntary basis or as most of the cases, it is a requirement by company policy or international regulation or by law (ETSI 2016). Standards play a significant role in the area of information and communication technology by addressing key requirements for interconnection and interoperability. They are also important to ensure safety, reliability, and environmental care. Standards offer a number of benefits for business, community, and individual. One of the key benefits is to ease the understanding and management of particular area of interest by various stakeholders. Standards are very important for business and they often adopt them for situations such as; business needs and requirements, regulators, and mandates. Al-Ahmad and Mohammad (2012) argued that the main reasons behind standards adoption is to meet the needs of businesses in certain areas due to the absence of expertise which hinders the establishment of proprietary standards based on their staff competencies.

8.2 Embedded Systems in CPS Standards

Cyber-physical system (CPS) is a combination of physical, sensing, computational, control, and networking elements (Ali et al. 2015). A set of sensors allow these systems to sense the physical environment and react accordingly through control and actuator modules with a physical change in the system. Applications of CPS can be seen in aviation, defense, energy distribution, and health sectors (Rho et al. 2016).

The foundation of the CPS was laid by the embedded systems. Marwedel (2010) discussed the relationship between CPS and embedded systems and how embedded systems are the backbone of the CPS, he also discussed the CPS design using embedded systems and provided some specifications on the requirements, challenges, constraints, applications, software coding, and all other concepts related to the CPS design. CPS is built on research from the domain of embedded systems and sensor network Parvin et al. (2013). While Lee (2010) described CPS as an intersection of cyber and physical realms rather than just a union. On the other hand, Broy et al. (2012), Broy and Schmidt (2014) defined CPS as "the integration of embedded systems with global networks such as the Internet" and the first step toward CPS is identified as networked embedded systems. Some scholars have listed other technologies like Internet as business, RFID, Semantic Web, and applications like Android, Firefox, etc., as the driving force toward CPS. While Magureanu et al. (2013) termed CPS as "massively distributed heterogeneous embedded systems linked through wired/wireless connections" and later as "embedded distributed systems".

Thus, it is important to mention the currently available standards established under the umbrella of embedded systems in CPS from organizational, business, technical, and societal perspective. In addition to those standards that are specifically designed for CPS. It is noted that currently, there is quite a number of organizations responsible of studying and issuing standards for several aspects of the electrical and electronic domains. Some of these organizations are as follows:

- International Electrotechnical Commission (IEC),
- International Organization for Standardization (ISO),
- American National Standard Institute (ANSI),
- International Telecommunication Union (ITU),
- Institute of Electrical and Electronics Engineers (IEEE),
- National Institute of Standards and Technology (NIST),
- Association Connecting Electronics Industries (IPC), and
- International Society of Automotive Engineers (SAE).

The above mentioned standards cover various aspects like hardware installations, software languages, testing devices, design and modeling, protection and security. On the other hand, there are standards that are released as per countries/nations perspective, since most of the countries around the world have either adopted standards from the international organizations listed above with few modifications or created their own standards for specific purposes. Some of these national organizations are as follows:

- Canadian Standards Association (CSA),
- SysAdmin, Audit, Networking and Security Institute Standards (SANS),
- Standardization Administration of China,
- Korea Industrial Standard,
- Japanese Standards Association,
- Russian National Standards,
- The British Standards Institution (BS),
- European Telecommunications Standards Institute,
- German Institute for Standardization,
- Association Francaise de Normalization Publication,
- Netherlands Standardization Institute (NEN),
- Dansk Standards (DS),
- The Spanish Association for Standardization and Certification and Standards Australia (AS), etc.

Table 8.1 provides a list of the standards related to embedded systems and CPS.

It is evident from the table above that majority of the organization have prepared standards based on their domain of information technology. Some of them are national standards (China, Japan, and Russia). There are also some domains that are redundant in certain standards, for example, the IPC and the IEC have their own

Table 8.1 Embedded systems in CPS standards

	Domain	Standard	Year released	Definitions	Regional adoption
ANSI/ SCTE	Cable TV	24 (1–23)	2009– 2013	Protocols, Security, Application	CSA, AS
	Fiber optic cable	165 (1-21)	2009	Framework, Installation, Security, Protocol, Management	
	Embedded cable modem	107	2009	Model, Requirements, Testing	
ISO/ IEC	Biometrics	29,164	2011 and 2013	Framework, Coding, Security, Management	ANSI, SANS, NEN
	Software engineering	25,051	2006 and 2014	Product quality requirements, Functionality, Reliability, Usability, Efficiency, Maintainability, Portability, and Testing	
	Computer graphics and image processing	180xx	2006– 2014	Programming language, Encoding, and Decoding	ANSI, NEN, DS, AS
		1977x	2008– 2015		
	Mixed and augmented reality	1852x	TBA	Model, Physical sensors, and Representation	
	Chip design (High speed I/O)	18,372	2004	Specifications, Model, Protocols, Guidelines	NEN (2005)
	Complex embedded systems	9496:	2003 and 2010	Programming Language CHILL, Models and classes, Input/Output, Structure, Specification, and Syntax	ANSI
IPC	Printed circuit board	601 (1-8), 9151, 9194, 9199, 9252, 9631, 9641, 9691	1999– 2013	Specification, Qualification and Performance, Reliability Benchmark,	

(continued)

Table 8.1 (continued)

Domain		Standard	Year released	Definitions	Regional adoption
				Guidelines, Quality, Testing, and Simulation	
	Embedded technologies	2316, 4811, 4821, 7091	2006–2007	Design Guide, and Specifications	
	Embedded passives devices	EMBPASWP309	2009	Implementation, Benefits, and Costs	
	Embedded components packaging	709 (2-5)	2009–2013	Design and Assembly Process Implementation	
IEC	Printed circuit board	61,188 (1-7)	1997–2009	Design and Use	
		61,189 (1-11)	1997–2015	Testing	
	System on chip	62,528	2007	Testing	BS, NEN
IEEE	Embedded core-based integrated circuits	1500	2009	Testability methods	
	Instrumentation in semiconductor devices	1687	2014	Access and Control	
	Information technology	1003 (1-26)	1991–2013	Testing, operating system, Interface, Language, and Application	
	Digital systems	1450 (1-6)	1999–2014	Standard Test Interface Language	

(continued)

Table 8.1 (continued)

	Domain	Standard	Year released	Definitions	Regional adoption
SAE	Distributed embedded systems	J2356	2007	Architectural model and Graphical techniques	
	Power	AS 90335A and MIL DTL 32385	2011	Connectors, receptacles, Plugs Adapters, PCB, panels	
	Embedded and real-time systems software	AS 5506	2012	Architecture Analysis and Design Language (AADL)	
	Vehicular embedded system software	J2640	2008	Best Practices for programming, hardware/software interface, multi-threaded system best practices, and verification criteria	
AIAG	Complex and embedded systems	PMCE	2008	Project management, Product integrity, and Program quality	
SA China	Smart card	GB/T 20276	2006	Information Security and embedded software	
	Embedded software	GB/T 28169	2011	C language coding, testing, and application	
		GB/T 28171	2011	Reliability Testing	
		GB/T 28172	2011	Quality Assurance Requirements	
	Embedded system life cycle	GB/T 28173	2011	Application management, engineering process, and implementation framework	
JSA (Japan)	Embedded system development	JIS × 0180	2011	Frameworks and guideline	
RNS Russia	Embedded system software	GOST R 51904	2002	Requirements, Documentation, and Development of embedded systems	

standards for the PCB technology. It is obvious that there is no single and combined standard for each domain of technology while some other standards are on constant revisions and updating in order to add the latest innovations in IT. Therefore, it is often challenging for organizations and customers to make an effective choice out of these varieties of standards.

8.3 CPS Standards

The increasing growth in CPS domain calls for a set of rules and procedures that can govern the platform to insure uniformity across the products and services. The presence of a set of standards would aid in achieving uniformity and would allow for seamless integration of CPS with mainstream infrastructures. Due to the nature of CPS for being a result of a cross-disciplinary approaches of engineering, information, and communication technologies (Shi et al. 2011), it makes it an interesting subject of standards. In this chapter, we studied different standards pertaining to CPS and how each of these standards, represented by their issuing body, addresses various aspects of CPS. The chapter reviews the relevant standards and attempts to identify the gaps that these standards fail to cover. As per the findings, the standards that apply to CPS can be classified into three groups according to the domain they focus on: management, operational, and technical standards as shown in Fig. 8.1.

- The management standards aim to bridge the gap between business and technology. These standards encompass three key areas: IT governance, internal control and enterprise risk management, and service management. IT governance

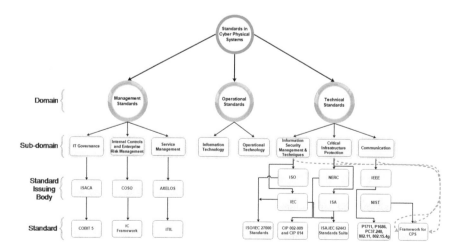

Fig. 8.1 CPS classification based on the domain focus

includes Control Objectives for Information and Related Technology (COBIT) developed by Information Systems Audit and Control Association (ISACA) that primarily focuses on the importance of IT governance and risk management in an organization (Tuttle and Vandervelde 2007). Committee of Sponsoring Organizations of the Tread way Commission (COSO) has also published an integrated control framework that deals mainly with the importance of internal controls and enterprise risk management (Laura and Michael 2003). The third component is service management where about the IT Infrastructure Library (ITIL) recommends several practices for IT and its alignment with the needs of the business (Axelos 2016).

- Operational standards on the other hand deal with guidelines that are prescribed for routine practices in technology. Operational standards can be decomposed into information technology (IT) and operational technology (OT).
- Technical standards that govern CPS can be classified into several domains based on the different dimensions that they address, such as information security management and techniques, critical infrastructure protection and communication. The standards that fall under each of these respective domains have been summarized in Fig. 8.1.

Information security management mainly deals with the safeguarding and security of assets within an organization and directs the codes of practice pertaining to security mechanisms and techniques. While *critical infrastructure protection* governs the safe keep of critical cyber-assets of an organization. It performs critical electronic system functions and the failure of which would drastically affect the reliability of operations. Adherence to the *communication standards* helps to maintain interoperability among platforms and devices by establishing generally accepted protocols as a baseline. The following sections will present a number of standards categorized under technical standards as illustrated in Fig. 8.1.

A number of international organizations and consortiums have tried to address the call for establishing generally accepted protocols. The ISO, North American Electricity Reliability Corporation (NERC) and the IEEE were key contributors in this regard. Under the joint efforts of both the ISO and the IEC, the ISO 27000 family of standards were defined to secure information assets. The ISO 27001 has been the best known in the industry for outlining a systematic approach to manage sensitive corporate information securely (ISO 2013a). The standard addresses requirements for an information security management system (ISMS). Another standard that belongs to same group of standards is ISO/IEC 27002:2013 that caters for control factors that determine the organization's information security risk environment(s) (ISO 2013b). Furthermore, the ISO/IEC 27003:2010 focuses on the design and implementation phases of a successful ISMS (ISO 2010). The ISO/IEC 27004:2009 standard also forms a part of the family, which guides in the use of metrics to assess the effectiveness of an ISMS (ISO 2009).

Standards from NERC mainly address the critical infrastructure protection domain. The critical infrastructure protection cybersecurity standard issued by

NERC views an organization's information assets as an essential part of the business process. It introduces a set of controls from a managerial perspective in the protection of the cyber-assets. The currently active standards that range from CIP 002 through CIP 009 cover particular aspects of cybersecurity. These aspects range from the identification of critical cyber-assets and controls to training up for incident reporting and recovery plans for these assets in case of any failure. While CIP 014 deals with the aspect of physical security in the sense that physical damage to transmission stations or substations and their associated control centers could result in instability due to a possible physical attack (NERC 2008).

On the other hand, ISA and the IEC have come together to draft the ISA. IEC-62443 series of standards that defines a secure implementation procedures for industrial automation and control systems (IACS) (IEC 2013; ISA 2009). The IEEE proposed some guidelines that govern several aspects pertaining to communication. Standards developed by IEEE address several elements concerning communication security, networking specifications, standards for intelligent electronic devices, and cybersecurity requirements at substations (IEEE 2010, 2014). Several aspects of CPS have been studied in detail by the NIST. The organization has devised a comprehensive framework that covers the entire dimensionality of CPS and the systems engineering process (PWG 2015). The purpose of this framework was to guide the designing and the building process for CPS and to act as a benchmark against which CPS could be verified. A common foundation such as this would aid in developing interoperable CPS that is safe and secure and follows a broadly adopted set of guidelines.

8.4 Regional CPS Standards

In addition to the standards discussed above, some other regional organizations come into mark their contribution to the growing field of CPS. European Committee for Standardization (CEN) that takes into account the interests of the countries within the European Union (EU) Jurisdiction brings together 33 European national standard issuing bodies. This effort is performed to establish mutual understanding on the specifications and other technical information that applies to products, materials, services, and processes (CEN 2016a, b). Another common regional standardization body is the European Telecommunications Standards Institute (ETSI) that publishes standards that are officially responsible for standardization across the continent. Although an officially recognized standardization body under the EU, their standards in the field of mobile telephony and smart grids among others have been globally applicable standards. They house over 800 member organizations from 66 countries (ETSI 2016).

The member states of the Gulf Cooperation Council (GCC) and Yemen are also subject to their regional standardization organization known as the GCC Standardization Organization (GSO) (Cerna 2015, GSO). The major standards which are applicable to CPS that are issued by the GSO are listed in Table 8.1.

As an organization still in its infancy, it is said to review existing global standards and adopt them as they deem fit, for those domains that the organization does not already address. The Directorate General of specification and standards specify guidelines those are specific to the Sultanate of Oman (ISO 2016).

8.5 CPS Standards: Comparative Analysis

There have been major efforts by the published CPS standards to address issues and concerns in CPS. It is also observed that most of the standards discussed in earlier sections of this chapter highlight concerns based on domains such as techniques and management of information security, protection of critical infrastructure elements, and practices governing communication. In addition to addressing these major issues pertaining to the platform, it is essential that special focus be laid on cybersecurity. Despite the fact of CPS criticality, there grows a need for a special attention to be laid on cybersecurity due to the increases in cybercrimes rates and given the continuous flow of information through CPS which makes it subject to security risk as opposed to stand-alone platforms (Atzori et al. 2010; Gandhi 2012; Harrison and Pagliery 2015).

Table 8.2 lists the standards that are in relations to CPS with respect to the specific areas that each of these standards focuses on. Given the area of interest that each of the respective standard issuing bodies primarily works on, most of the standards have been developed based on a single-dimension model solely addressing their particular field of work. These standards broadly address issues relating to their domain very systematically. Although these standards have been developed based on comprehensive studies and research, most of the standards fail to address the requirements that a multidisciplinary platform that CPS entails.

The organizational aspects of the management of CPSs are very well addressed by bodies such as ISACA, COSO, and AXELOS. They address best practices, policies, and procedures for areas such as IT governance, internal controls, enterprise risk management, and service management are covered in depth by each of these distinct frameworks. To have an extensive grip on the managerial aspects of CPS, it is essential that all of the above be implemented in union with each other.

Operational standards, as discussed earlier in this chapter, are divided into information and operation technology. Best practices and suggestions with the operational aspects of CPS are particularly dependent on the organization setup that the infrastructure is present in. It is also to be noted that there exists a shortage of standard issuing bodies that address issues under this domain.

The manner in which the standards fall short from well-known organization such as the ISO, IEC, NERC, ISA, and IEEE is in terms of competency in the isolated nature of addressing each of the elements separately. For instance, the ISO 27000 standards suite almost comprehensively covers the areas of information security management and techniques. Communication protocols on the other hand are addressed by the different IEEE standards. The issue here is that CPS is a vast

Table 8.2 Comparison of standards applicable to CPS

Domain standard (Issuing body)	Management				Technical		
	IT governance	Internal controls	Enterprise risk management	Service management	Information security management and techniques	Critical infrastructure protection	Communication
COBIT 5 (ISACA)	✓						
IC Framework (COSO)		✓	✓				
ITIL (AXELOS)				✓			
ISO/IEC 27000 Suite (ISO/IEC)					✓		
CIP 002-009 and CIP 014 (NERC)						✓	
ISA.IEC 62443 Suite (ISA/IEC)						✓	
P1711, P1686, PC37.240, 802.11 and 802.15.4g (IEEE)							✓
Framework for CPS (NIST)					✓	✓	✓

platform of which the areas that the above standards cover are only a subset. The broad scope of CPS requires a comprehensive and structured approach that encompasses the varied range of concepts that the platform touches upon. This is where the framework for CPS drafted by the NIST has a significant leverage over its peers (PWG 2015).

NIST through their Public Working Group (PWG) widely covered the different aspects and domains pertaining CPS in their framework for the platform (PWG 2015). The framework aims to allow for a comprehensive analysis for CPS by establishing a common body of knowledge for the industry to work on. With the CPS being a multifaceted platform, it is critical that each of its elements be assessed in detail. As opposed to the other standards in Table 8.2, NIST made remarkable attempts to identify all the major components of CPS in their framework and has discussed the platform based on different domains, facets, aspects, and concerns. Although the framework is at its level of infancy, it has still taken into consideration the foundational concepts of CPS along with the cross-cutting concerns with respect to several of its domains. In the context of an adoption of a well-drafted standard suite(s) for CPS is concerned, the authors believe that the framework drafted by NIST does absolute justice in addressing the concerns which are most vital in the success of the architecture. However, there are some important features that need to be considered in the adoption of CPS. For instance, cybersecurity is a vital concern for digital assets today and the cyber element of CPS makes the field also to be vulnerable to a certain amount of risks, given its shared interests.

8.6   Conclusion

Cyber-physical system (CPS) is a combination of physical, sensing, computational, control, and networking elements. Due to the increasing growth in CPS domain, there is a need for rules and procedures that can govern the platform to insure uniformity across the products and services. The presence of a set of standards would aid in achieving uniformity and would allow for seamless integration of CPS with mainstream infrastructures. There are a number of standards that are applied in CPS domain and divided into three main categories; management, operational, and technical. Although there have been major efforts by the published CPS standards to address issues and concerns in CPS, there is still a need for set of rules and procedures that deal with CPS security.

References

Al-Ahmad, W., & Mohammad, B. (2012). Can a single security framework address information security risks adequately? *International Journal of Digital Information and Wireless Communications, 2,* 222–230.

Ali, S., Anwar, R. W., & Hussain, O. K. (2015). Cyber security for cyber physical systems: A trust-based approach. *Journal of Theoretical and Applied Information Technology, 71,* 144–145.

Atzori, L., Iera, A., & Morabito, G. (2010). The internet of things: A survey. *Computer Networks, 54,* 2787–2805.

Australia, S. (2015). *What is a standard?* [Online]. Standards Australia. Available: http://www.standards.org.au/StandardsDevelopment/What_is_a_Standard/Pages/default.aspx. Accessed September 28, 2015.

AXELOS. (2016). *What is ITIL® best practice?* [Online]. Axelos. Available: https://www.axelos.com/best-practice-solutions/itil/what-is-itil. Accessed January 25, 2016.

Broy, M., Cengarle, M. V., & Geisberger, E. (2012). Cyber-physical systems: Imminent challenges. In *Large-scale complex IT systems. Development, operation and management.* Springer.

Broy, M., & Schmidt, A. (2014). Challenges in engineering cyber-physical systems. *Computer,* 70–72.

CEN. (2016a). *European standardization* [Online]. European Committee for Standardization. Available: http://www.cen.eu/you/EuropeanStandardization/Pages/default.aspx. Accessed January 20, 2016.

CEN. (2016b). *Who we are* [Online]. European Committee for Standardization. Available: http://www.cen.eu/about/Pages/default.aspx. Accessed January 22, 2016.

Cerna, F. D. L. (2015). *How green are we?* [Online]. Thomson Reuters Zawya. Available: https://www.zawya.com/story/How_green_are_we-ZAWYA20150513074331/. Accessed January 18, 2016.

Commission, I. E. (2016), International Standards and Conformity Assessment for all electrical, electronic and related technologoes. http://www.iec.ch/standardsdev/publications/is.htm, Accessed on: November 28th, 2016.

ETSI. (2016). *Our standards* [Online]. European Telecommunication Standardization Institute. Available: http://www.etsi.org/standards. Accessed January 25, 2016.

Gandhi, K. (2012). An overview study on cyber crimes in internet. *Journal of Information Engineering and Applications, 2.*

GSO. *Standards* [Online]. GCC Standardization Organisation. Available: http://www.gso.org.sa/gso-website/gso-website/activities/standards. Accessed January 20, 2016.

Harrison, V., & Pagliery, J. (2015). *Nearly 1 million new malware threats released every day* [Online]. CNN Money. Available: http://money.cnn.com/2015/04/14/technology/security/cyber-attack-hacks-security/. Accessed October 13, 2015.

IEC. (2013). *Industrial communication networks—Network and system security* [Online]. International Electrotechnical Commission. Available: https://webstore.iec.ch/publication/7033. Accessed March 9, 2016.

IEC. (2015). *International standards (IS)* [Online]. International Electrotechnical Commission. Available: http://www.iec.ch/standardsdev/publications/is.htm. Accessed September 28, 2015.

IEEE. (2010). *IEEE SA—P1711—Standard for a cryptographic protocol for cyber security of substation serial links* [Online]. Institute of Electrical and Electronics Engineers. Available: https://standards.ieee.org/develop/project/1711.html. Accessed March 9, 2016.

IEEE. (2014). *IEEE SA C37.240-2014—IEEE Standard cyber security requirements for substation automation, protection, and control systems* [Online]. Institute of Electrical and Electronic Engineers. Available: https://standards.ieee.org/findstds/standard/C37.240-2014.html. Accessed March 9, 2016.

ISA. (2009). *ISA99, Industrial automation and control systems security* [Online]. International Society for Automation. Available: https://www.isa.org/isa99/. Accessed March 9, 2016.

ISO. *Standards* [Online]. International Organization for Standardization. Available: http://www.iso.org/iso/home/standards.htm. Accessed January 14, 2016.

ISO. (2009). *ISO/IEC 27004:2009 Information technology—Security techniques—Information security management—Measurement* [Online]. International Organization for Standardization. Available: http://www.iso.org/iso/catalogue_detail?csnumber=42106. Accessed January 20, 2016.

ISO. (2010). *ISO/IEC 27003:2010 Information technology—Security techniques—Information security management system implementation guidance* [Online]. International Organization for Standardization. Available: http://www.iso.org/iso/catalogue_detail?csnumber=42105. Accessed January 18, 2016.

ISO. (2013a). *ISO/IEC 27001—Information security management* [Online]. International Organization for Standardization. Available: http://www.iso.org/iso/home/standards/management-standards/iso27001.htm. Accessed January 16, 2016.

ISO. (2013b). *ISO/IEC 27002:2013 Information technology—Security techniques—Code of practice for information security controls* [Online]. International Organization for Standardization. Available: http://www.iso.org/iso/catalogue_detail?csnumber=54533. Accessed January 20, 2016.

ISO. (2016). *ISO member body* [Online]. International Organisation for Standardization. Available: http://www.iso.org/iso/about/iso_members/iso_member_body.htm?member_id=2007. Accessed January 24, 2016.

Laura, F. S., & Michael, P. (2003). Risk management: The reinvention of internal control and the changing role of internal auditnull. *Accounting, Auditing & Accountability Journal, 16,* 640–661.

Lee, E. A. (2010). CPS foundations. In *Proceedings of the 47th Design Automation Conference* (pp. 737–742). ACM.

Magureanu, G., Gavrilescu, M., & Pescaru, D. (2013). Validation of static properties in unified modeling language models for cyber physical systems. *Journal of Zhejiang University SCIENCE C, 14,* 332–346.

Marwedel, P. (2010). *Embedded system design: Embedded systems foundations of cyber-physical systems.* Springer Science & Business Media.

NERC. (2008). *CIP Standards* [Online]. North American Electric Reliability Corporation. Available: http://www.nerc.com/pa/Stand/Pages/CIPStandards.aspx. Accessed March 9, 2016.

Parvin, S., Hussain, F. K., Hussain, O. K., Thein, T., & Park, J. S. (2013). Multi-cyber framework for availability enhancement of cyber physical systems. *Computing, 95,* 927–948.

PWG, C. (2015). *CPS PWG draft cyber-physical systems (CPS) framework* [Online]. National Institute of Standards and Technology. Available: https://pages.nist.gov/cpspwg/. Accessed March 9, 2016.

Rho, S., Vasilakos, A. V., & Chen, W. (2016). Cyber physical systems technologies and applications. *Future Generation Computer Systems, 56,* 436–437.

Shi, J., Wan, J., Yan, H., & Suo, H. (2011). A survey of cyber-physical systems. In *2011 International Conference on Wireless Communications and Signal Processing (WCSP)* (pp. 1–6). IEEE.

Tuttle, B., & Vandervelde, S. D. (2007). An empirical examination of CobiT as an internal control framework for information technology. *International Journal of Accounting Information Systems, 8,* 240–263.

Printed in the United States
By Bookmasters